Fundamentals
of
Descriptive Statistics

Zealure C. Holcomb

Routledge
Taylor & Francis Group

LONDON AND NEW YORK

First published 1998 by Pyrczak Publishing.

Published 2017 by Routledge
2 Park Square, Milton Park, Abingdon, Oxon OX14 4RN
711 Third Avenue, New York, NY 10017, USA

Routledge is an imprint of the Taylor & Francis Group, an informa business

Cover design by Mario Sanchez.

ISBN-13: 978-1-884-58505-0 (pbk)

Contents

Introduction   *v*

1. Introduction to Statistics   1

2. Percentages and Bar Graphs   7

3. Shapes of Distributions   13

4. The Mean: The Most Popular Average   21

5. The Median and Mode: Alternatives to the Mean   27

6. The Standard Deviation: First Cousin of the Mean   33

7. The Range, Quartiles, and Interquartile Range   43

8. Standard Scores   49

9. Scattergrams and the Pearson r 57

10. The Coefficient of Determination   71

11. Linear Regression   77

12. Glossary of Descriptive Statistics   87

Appendix A Formula for the Interpolated Median   91

Appendix B Computational Formula for the Standard Deviation   93

Appendix C Computational Formula for the Pearson r 95

Index   97

NOTES

Introduction

When there are large amounts of data that need to be interpreted, descriptive statistics are used to organize and summarize them. For example, suppose a teacher administered a standardized test to the 60 fifth graders in a school and obtained both a reading and math score for each student. The scores can be analyzed with descriptive statistics to answer questions such as "What is the typical performance of the students on each test?" or "On the whole, are these students better at reading or at math?" or "How does this group compare to the national norm group?" You will learn how to answer such questions in this book.

Of course, teachers are not the only professionals who need descriptive statistics. Psychologists and social workers, for example, gather data on the clients they work with. Marketing researchers gather data on consumers' reactions to products and services. Administrators in all types of organizations continuously need to direct the collection and interpretation of data. The principles and techniques described in this book apply equally well to all these fields.

What are *descriptive statistics*? They are tools that help us organize and summarize data. Examples that you are already familiar with are graphs, percentages, and averages. We'll explore the use of these and other descriptive tools in this book.

A Note on Computations

The widespread availability of computers has greatly reduced the computational burden on students of statistics. With a statistics program, you can enter data, select the statistics you wish to have computed, and get the answers without examining the formulas. Yet, while you are learning the meanings of statistics, consideration of computational formulas will help you develop an understanding of those statistics. Thus, you will find a number of formulas discussed throughout this book, and the end-of-chapter exercises provide an opportunity for you to practice their use with very small data sets.

If math makes you anxious, you will be glad to learn that none of the formulas in this book requires more than a very elementary knowledge of math. Performing the steps carefully as specified by the formulas and in order is all that is required to get correct answers.

Acknowledgments

Dr. Richard Rasor of the American River College and Dr. Robert Morman of the California State University – Los Angeles reviewed the first draft of this book and provided many useful suggestions. Errors and omissions, of course, remain the responsibility of the author.

Chapter 1

Introduction to Statistics

Here's an example that illustrates the need for statistics:

Researchers studied the emotional health of 58 12- to 19-year-old foster care adolescent males who reside in a group home in Virginia. Emotional health was measured by having the adolescents respond to statements such as "Sometimes I feel so ashamed of myself that I just want to hide in a corner and cry" and "I often blame myself even when I am not at fault." Responses were scored so that higher scores indicate better emotional health. The average score for the national standardization group was 50. These scores were obtained for the 58 foster care males:[1]

63	28	59	4	59	37	30	37	43	53
14	46	3	39	47	48	46	32	46	14
60	47	67	64	66	55	53	21	42	48
50	55	9	52	44	29	54	73	55	47
32	47	31	59	9	59	52	52	39	41
27	42	65	62	36	42	42	34		

As you can see, the researchers have a problem: how can these scores be organized and summarized so that the researchers can concisely discuss the overall emotional health of this group of adolescent males? The problem is compounded by the fact that the researchers measured ten additional variables such as sexual attitudes and impulse control, yielding a total of 638 scores for the 58 adolescents. Clearly, such a large number of scores needs to be summarized in order to produce an intelligible description of the results. We will return to the scores in this example at several points throughout this book to illustrate how descriptive statistics are used to organize and summarize data.

---

[1]Data supplied by Dr. Susan B. Lyman, School of Social Welfare, State University of New York at Albany. For more information on this topic, see Lyman, S. B., & Bird, G. W. (1996). A closer look at self-image in male foster care adolescents. *Social Work, 41*, 85-96.

Populations and Samples

A *population* is the group of interest to a researcher. It may be large, such as all licensed clinical psychologists in the United States, or it may be small, such as all clients served by a social worker. In the example we started with, the population consists of the small number of boys who reside in a foster care home.

When a population is large, we can draw a *sample* from it, study those in the sample, and generalize the results to the population (that is, infer that what is true of the sample is also true of the population).

Descriptive statistics are used to organize and summarize data whether they come from studies of populations or samples.[2] However, another type of statistics called *inferential statistics* is needed for making generalizations from samples to populations. For example, if a poll of 1,000 registered voters indicates that 55% approve of how the President is handling the economy, inferential statistics can be used to compute a margin of error, which is an allowance for the possible fluctuations due to sampling. Thus, if the margin of error is 4 percentage points, we can be confident that the true percentage of the population who approve is between 51% and 59% (that is, 55% plus and minus 4%). Although inferential statistics perform an important function when we have sampled, they are *not* needed when we analyze the data of entire populations since there is no sampling error when we do not sample.

Note that each researcher defines a population of interest when planning research. For example, one researcher's population might be all registered nurses in California while another researcher's population might be all registered nurses employed by a specific hospital. Thus, a population may be large or small, depending on the researcher's interest.

Studies of entire populations are more common than you might realize. They are very common in institutional settings such as schools, hospitals, prisons, the military, and corporations. For example, a teacher might measure the reading ability of all third graders in a classroom, and an administrator might measure the same variable for all third graders in the school district. Studies of entire populations are also common when individuals participate in a program that requires them to provide certain information.

---

[2]Statisticians tell us that when a value is based on the study of a population, it should be called a *parameter*. When it is based on a sample, it should be called a *statistic*. In practice, applied researchers usually call both values simply *statistics*.

Examples are people receiving welfare, the clients of a free health clinic, seniors on Medicare, and so on. Thus, those of you who are preparing for jobs in institutional settings or for jobs in which you will be delivering program services will probably find yourselves analyzing population data on the job. Also, you will very likely be required to read various reports on the job that contain descriptive statistics pertaining to the populations you serve. This book will give you a solid foundation in how to perform these tasks.

Scales of Measurement

Researchers measure with a wide variety of measuring tools (known as *instruments*) such as paper-and-pencil tests, interviews, direct observations of behavior, and self-report questionnaires. The data derived from such instruments may be classified according to *scales of measurement*. We will consider the *nominal*, *ordinal*, and *equal interval* scales, which will be useful in our discussions in later chapters.

At the *nominal* level, we classify individuals with words instead of numbers. For example, we might ask people to name their HIV status using these categories: Positive, Negative, and Don't Know; or we might ask people to name their political affiliation with these categories: Democrat, Republican, and Other. Thus, nominal data may be thought of as "naming data." This type of data is also referred to as *categorical data*.

When we classify individuals by putting them in rank order, we obtain *ordinal* data. For example, if we rank the students in a class according to their height by giving a rank of 1 to the tallest student, 2 to the next tallest student, and so on, the ranks constitute ordinal data. A fundamental weakness of ordinal data is that they fail to indicate by how much the individuals differ from each other. For example, the student with a rank of 1 on height may be only one-half inch taller than the student with a rank of 2, while the student with a rank of 2 may be three inches taller than the student with a rank of 3. Because of this weakness, we try to limit our use of ordinal data, preferring instead data obtained at the next level.

At the *equal interval* level, we classify individuals along a numerical continuum that has equal distances among the values. When you examine a ruler, you see equal intervals. The distance between the numbers 1 and 2 is the same as the distance between 2 and 3, and between 3 and 4, and so on. Researchers usually regard test scores as being equal interval. For example, scores on a multiple-choice achievement test are usually assumed to be equal interval. That is, the difference between having 0 right and 1 right on the test is assumed to represent the same amount of achievement as the difference between having 1 right and having 2 right. Likewise, scores on objective measures of attitude, temperament, and personality are usually assumed to be equal interval. For example, if we measure the job satisfaction of social workers using 20 statements to which they respond "true" or "false," we can count how many statements were responded to favorably by each respondent, yielding scores that range from 0 to 20. Likewise, we could let the social workers respond on a 5-point scale from "strongly agree" to "strongly disagree" and obtain a total score for each respondent. Such scores are usually assumed to have equal intervals among them.[3]

Since the three levels of measurement are referred to at various points in this book, it's a good idea to pause now and master them. Remember:

> NOMINAL → naming data
> ORDINAL → rank order data
> EQUAL INTERVAL → scores with equal intervals

The next chapter describes how to analyze nominal data.

EXERCISE FOR CHAPTER 1

Factual Questions

1. A county sheriff is interested in determining the HIV status of all inmates in the county jail. If he examines the records of all inmates to determine the number who reported that they are HIV+, the sheriff is studying a
   A. population. B. sample.

---

[3]Technically, there are two classifications for equal interval data. When there is a true zero point on a scale, such as the 0 on a ruler, it is a *ratio* scale. When there is no true zero point, it is an *interval* scale. For example, you could get a zero score on an intelligence test, but this does not mean that you truly have zero intelligence. Thus, while intelligence test scores are assumed to have equal intervals, they are not called ratio. This distinction has no implications for the types of statistical analyses widely used by applied researchers. Thus, in this book, the *ratio* and *interval* scales are grouped together as the *equal interval* scale.

2. Descriptive statistics are used to organize and summarize data obtained from
   A. samples only. B. populations only.
   C. both samples and populations.

3. A city health official examined the coroner's records to determine how many men and women in a city died of sudden cardiac arrest during a recent year. To the best of her knowledge, the records are complete and accurate for the population. Does she need to use inferential statistics to interpret the data?
   A. yes B. no

4. Are all populations large?
   A. yes B. no

5. If the science projects of the five finalists in a contest in a school district are ranked from 1 to 5, which scale of measurement is being used?
   A. nominal B. ordinal C. equal interval

6. If respondents in a survey are asked to name their state of residence, which scale of measurement is being used?
   A. nominal B. ordinal C. equal interval

7. Scores obtained by using an objective multiple-choice test are usually classified as belonging to which scale of measurement?
   A. nominal B. ordinal C. equal interval

Questions for Discussion

8. Name a population that you might need to study on the job in either your current or future occupation. Would you be able to study it without sampling? Explain.

9. Name a variable other than those mentioned in this chapter that, when measured, would yield nominal data.

NOTES

Chapter 2

Percentages and Bar Graphs

As you recall from the previous chapter, nominal data are "naming data." For example, the foster care adolescents whom we considered in Chapter 1 were asked to name their race, and 12 said African American while 46 said White. Calculating the corresponding *percentages* is easy: divide the number of individuals that we are interested in by the total number, multiply by 100, and add a percent sign. Here's how to do it for the African American adolescents, keeping in mind that there are 12 African Americans (the part) out of a total of 58 adolescents (the whole):

$$\frac{part}{whole} = \frac{12}{58} = (.207)(100) = 20.7 = 21\%$$

For the Whites, the percentage is calculated as follows:

$$\frac{part}{whole} = \frac{46}{58} = (.793)(100) = 79.3 = 79\%$$

What do these percentages tell us? Simply that for every 100 participants, roughly 21 are African American and 79 are White. Of course, there were only 58 participants, so why are we using percentages to convert to a base of 100? Because it allows us to compare groups of unequal size. For example, according to the United States Census, 23% of foster care adolescents in the nation are African American, 60% are White, 10% are Hispanic, 5% are classified as "other," and 2% are of unknown ethnicity. Thus, by comparing percentages, we can see that in the foster care home of interest there is a smaller percentage of African Americans (21%) than in homes nationally (23%). Also, there is a larger percentage of Whites (79%) in the home of interest than in homes nationally (60%).

As a partial check on your work, you should sum the percentages to see if it is *approximately* 100%. Thus, 21% + 79% = 100% for the local group, and 23% + 60% + 10% + 5% + 2% = 100% for the national one. Note that the sum will not always be exactly 100% because of rounding. An example is shown in Table 2.1. Although the arithmetic is correct, the

Table 2.1 Political Affiliation of Registered Voters in Uptown City

Political Affiliation	Number	Calculation	Percentage[1]
Democrat	322	322/726 = .4435	44%
Republican	288	288/726 = .3967	40%
Independent	90	90/726 = .1240	12%
Reform	16	16/726 = .0220	2%
Other	10	10/726 = .0138	1%
Total	726		99%

[1]Percentages do not sum to 100% because of rounding.

sum is only 99%. This occurred because, in this particular example, we happened to round down more often than we rounded up. Of course, there will be other cases where we will round up more often than down, in which case the sum may be slightly more than 100%. When the sum is not exactly 100%, it's a good idea to point this out with a footnote such as the one in Table 2.1.

The percentages in Table 2.1 have been rounded to whole numbers, which is common. However, some researchers report percentages to one or two decimal places in academic journals.

Table 2.1 is our first table, so let's pause to notice several features. First, it is called a *table* because statistical tables have organized sets of values (as opposed to a statistical drawing or *figure* such as a bar graph, which we'll discuss in the next section of this chapter). Second, the table has a number (2.1), which allows us to refer to the table by its number. In this case, the 2 stands for Chapter 2 and the 1 means that it is the first table in the chapter. Third, it has a title (also known as a *caption*), which concisely describes its content. Solid horizontal lines (called *rules*) are used to set off the table from the text.

Also note that all tables should be referred to in the text of the report in which they are included. If their contents are not worth even a brief mention, they are probably superfluous.

Finally, you should always report the total number of cases on which the percentages are based, which is done in the bottom row of Table 2.1.

Converting from a percentage to the number of cases

Suppose you read that 11% of the students at Braintrust University are psychology majors. If you know that there are 6,345 students at the university, here's how you calculate the number who are psychology majors:

> Move the decimal point in the percentage two places to the left and multiply by the total number. Thus, 11% becomes .11. Multiplying, we get .11 x 6,345 = 697.95, which rounds to 698.[1]

Bar Graphs

Bar graphs can be used to represent percentages. Figure 2.1 is a bar graph showing the percentages from Table 2.1.[2]

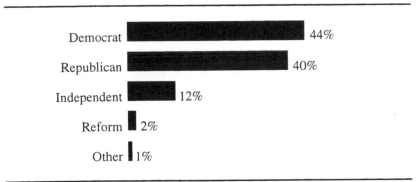

Figure 2.1   Political Affiliation of Registered Voters in Uptown City

You can make a neat and accurate bar graph by using graph paper. Here's how: First, consider how wide the longest bar would be if you were to allow one space on the graph paper for each percentage point. For the example shown above, the bar for the Democrats would be 44 spaces wide. You can easily manipulate this width. For example, if you want the bar to be 50% larger, multiply 44 by 1.5, which yields 66; thus, you would use 66 spaces for the Democrats. If you wanted it to be only 75% as large, multiply it by .75, which yields 33; you would make the bar for the Democrats 33 spaces wide. Of course, whatever multiplier you use with the Democrats must be used with all the other categories.

---

[1] Of course, there can't be 697.95 students. We got this result because the percentage (11%) had been rounded from a percentage such as 11.00078. . .%.

[2] Note that one of the most popular style guides in the social and behavioral sciences, the *Publication Manual of the American Psychological Association*, suggests that table numbers and titles be placed above tables while figure numbers and captions be placed below figures. We will follow this convention throughout this book.

As you can see, you have a choice between presenting percentages in a table or a bar graph. Another choice is simply to give the percentages in the narrative of the report. The choice depends on the importance of the percentages in your research. If the percentages are crucial to your main points, use a table or bar graph to draw attention to them.

EXERCISE FOR CHAPTER 2

Factual Questions

1. Percentages are expressed with a base of
   A. 1   B. 10   C. 100   D. 1,000

2. Only 67 of the 90 nursing majors who graduated from Health College remained in their chosen occupation for more than two years. What percentage remained in the occupation?

3. At Mindful University 2,489 of the 4,689 students are receiving financial aid, and at Sharper University 4,201 of the 12,515 students are receiving aid. A larger percentage of students are receiving aid at which university?

4. Suppose you read that exactly 8% of the 11,549 residents of a town had sought some form of professional counseling during the past year. How many sought counseling?

5. A bar graph is an example of a
   A. statistical table.   B. statistical figure.

6. Draw a bar graph like that shown in this chapter on graph paper for the percentages shown in the box below, and place it in the box on the next page. Plan the graph in advance so that it fits neatly in the space.

> Percentages planning to vote for each candidate:
> Candidate Doe, 42%; Candidate Jones, 38%; Undecided, 20%.

Note to students:
Attach your bar graph for question 6 here.

Questions for Discussion

7. Do you believe that *tables* or *bar graphs* are more effective in drawing the readers' attention to percentages? Why?

8. Which scale of measurement is represented by the data you used in question 6? (See Chapter 1 to review scales of measurement.) Explain.

NOTES

Chapter 3

Shapes of Distributions

When analyzing *equal interval* data,[1] the first thing you should do is to examine the shape of the distribution. To do this, prepare a *frequency distribution*, such as the one in Table 3.1 for the emotional health scores of the foster care adolescents on page 1 of this book.

Table 3.1 Frequency Distribution of Scores on Page 1

Scores (X)	Tally marks	Frequencies (f)	Percentages* (%)
73-77	/	1	1.7
68-72		0	0.0
63-67	/////	5	8.6
58-62	///// /	6	10.3
53-57	///// /	6	10.3
48-52	///// /	6	10.3
43-47	///// ////	9	15.5
38-42	///// //	7	12.1
33-37	////	4	6.9
28-32	///// /	6	10.3
23-27	/	1	1.7
18-22	/	1	1.7
13-17	//	2	3.4
8-12	//	2	3.4
3-7	//	2	3.4
Total		58	99.6%

*Percentages do not sum to 100.0% because of rounding.

---

[1] See Chapter 1 to review the scales of measurement: nominal, ordinal, and equal interval.

As you can see, the scores listed in the first column have been grouped into what are called *score intervals*. (Note that the symbol for scores is *X*.) In this example, each score interval is 5 points wide; for example, the bottom interval from 3 to 7 covers these five values: 3, 4, 5, 6, 7. There are two tally marks for this score interval because two of the adolescents had scores in the interval. In the next column are the frequencies, whose symbol is *f.* In the last column are percentages for each interval. It's easy to compute these: just divide each frequency by the total number of cases and multiply by 100. For example, for the 3–7 interval, 2/58 = 0.034 x 100 = 3.4.

Examining Table 3.1, we can see that the interval with the largest number of adolescents is 43–47 and that the majority of the adolescents are within two intervals of the 43–47 interval. Then there is a scattering of adolescents above and below this middle area.

Here are some pointers for preparing frequency distributions:

1. Select an interval size that will yield about 15 intervals. (It's okay to have as few as 10 or as many as 20.) A simple way to do this is first to compute the range of the scores by subtracting the lowest score from the highest and adding 1. For the scores on page 1, the range is 73 – 3 = 70 + 1 = 71. Then try dividing various whole numbers into the range of 71 until you get an answer that is between 10 and 20. Here's how it works:

   71/2 = 35.5 (which indicates that an interval size of 2 would yield 36 intervals, which would be too many intervals),

   71/3 = 23.7 (which would still be too many intervals),

   71/4 = 17.5 (which would be an acceptable number of intervals)

   71/5 = 14.2 (which means we would have 15 intervals if we use an interval size of 5, which is just about perfect, so 5 becomes the interval size).[2]

2. Build the intervals from the bottom up, starting with the lowest score, which in our example is 3. Thus, the bottom interval is 3 to 7, which covers 5 points. The next interval is 8 to 12, and so on. Notice that the top interval is 73 to 77 even though the highest score is 73. This occurred because of the requirement that all intervals must be the same size.

---

[2] Note that even though 71/5 = 14.2, which would ordinarily round to 14, we get 15 intervals since the two-tenths indicates that we will get *more than* 14 intervals, which is 15.

3. Tally the scores, crossing off each score as you tally it on the frequency distribution. Delete the tally marks when you type your research report.

Polygons

We can see the shape of the distribution better if we build a *polygon* such as the one in Figure 3.1. This was constructed on graph paper by placing a dot for the percentage in each score interval (shown in Table 3.1) and then connecting the dots.[3] The polygon makes it easy to see how most of the adolescents scored and how high and low their scores go.

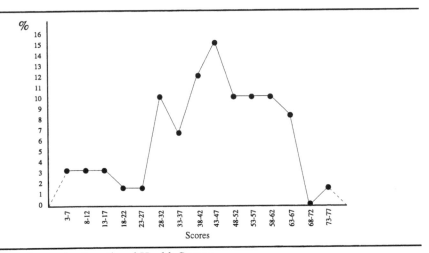

Figure 3.1 Polygon for Emotional Health Scores

When large populations are examined, polygons tend to be smooth; thus, they are often called *curves*. The most common type of curve is the bell-shaped curve, called the *normal curve*, which is shown in Figure 3.2. We find the normal curve all around us. For example, consider the heights of women. There are very few extremely tall women (thus, the curve is low at the high end) and there are extremely few very short women (thus, the curve is low at the low end). Most women are near middle or average (thus, the curve is high at the middle). Another example is the annual amount of rainfall in most areas. There is very little rainfall during a small

---

[3]Some researchers plot frequencies instead of percentages. However, when plotting polygons for two groups of unequal size on the same graph such as using a solid line for the 58 adolescent males and a dashed line for 77 males in another home, percentages should be used to make the two polygons comparable.

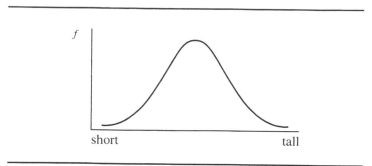

Figure 3.2 The Normal Curve

number of years, moderate amounts during most years, and tremendous amounts during a small number of years, which forms a normal curve.

A less common type of curve is a *skewed curve*. Figure 3.3 shows a curve that is *skewed to the right* (also called a curve with a *positive skew*). Income is skewed to the right because most people have relatively low incomes while a small percentage of people have very high incomes. Those with high incomes create a longer tail to the right. Because income is skewed to the right, many variables associated with income are also skewed to the right. For example, the distribution of housing prices is skewed to the right. Most houses cost relatively small amounts (which allows most families to own a house) while a small percentage of houses are very expensive, creating a tail to the right.

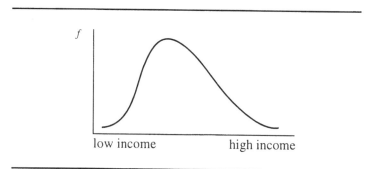

Figure 3.3 A Distribution Skewed to the Right (Positive Skew)

Some distributions are *skewed to the left* (also called a curve with a *negative skew*), which is shown in Figure 3.4. We might get a distribution skewed to the left, for example, if we tested a population of gifted students with a standardized math test. Most students should do very well because

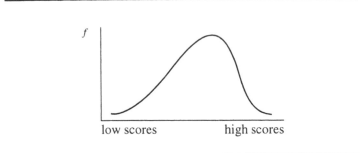

low scores   high scores

Figure 3.4 A Distribution Skewed to the Left (Negative Skew)

of their giftedness, but a scattering might not be terribly good at math (but gifted, instead, in other areas), creating a tail to the left.

You've probably already noticed that the types of skewness are named for the tails. When the longer tail points to the right, we say that a distribution is skewed to the right. When the longer tail points to the left, we say the distribution is skewed to the left.

So what about our curve for the emotional health scores of the adolescent foster care males that we saw in Figure 3.1? Is it normal or skewed? Well, it does have a high point near the middle with most of the males near this point, and it does drop off on both sides (but with a longer tail to the left than to the right). A good description might be "It's approximately normal with a slight tendency to be negatively skewed." As you'll learn in the following chapters, whether a curve is normal or skewed has important implications for the selection of other statistics. Most researchers consider a curve that is "approximately normal" as a normal curve.

An alternative to a polygon is a *histogram*. We build a histogram by drawing vertical bars to indicate the frequency for each score interval. Figure 3.5 shows a histogram of the emotional health scores. Comparing it with the polygon in Figure 3.1, you will see that they both convey the same information.

Presenting a polygon or histogram is a good way to describe a distribution in a research report. However, these statistical figures take up more space than other methods of describing distributions that we'll be exploring in later chapters. When space is limited, such as it usually is in academic journals, editors frown on including a large number of polygons or

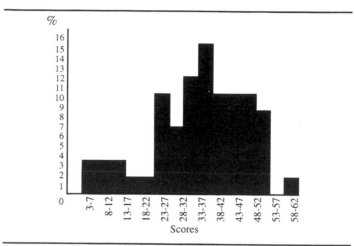

Figure 3.5 Histogram for Emotional Health Scores

histograms. When space is less limited, such as in a thesis or dissertation, a larger number might be included.

EXERCISE FOR CHAPTER 3

Factual Questions

1. According to Table 3.1 on page 15, how many adolescents had scores from 58 to 62?

2. According to Table 3.1, which score interval has the smallest percentage of respondents?

3. A frequency distribution should have about how many intervals?

4. Is a skewed distribution symmetrical?

5. What is another name for the bell-shaped curve?

6. What is another name for a skew to the left?

18

7. Do journal editors encourage researchers to include large numbers of polygons and histograms in their reports?

8. Is the curve shown below skewed?

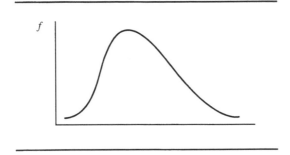

9. If your answer to question 8 is "yes," is it skewed to the right or skewed to the left?

10. Prepare a frequency distribution that shows the frequencies and percentages for the scores in the box below. Then draw a polygon based on the percentages. The scores are "number of days absent from school" for the population of 10th-grade boys in a small school in Canada.[4] Use an interval size of 7 and start the lowest interval with the score of 6.

26	38	28	20	16	28
36	90	111	21	20	26
25	6	14	50	25	36
66	80	19	60	22	41
30	9	28	50		

11. Examine the frequency distribution and polygon that you constructed for question 10. Is the distribution skewed? If yes, what type of skew does it have?

[4]Data supplied by Dr. T. F. McLaughlin, Department of Special Education, Gonzaga University, Spokane, WA.

Questions for Discussion

12. In question 10, you were told to "use an interval size of 7." Why do you think you were asked to use 7 instead of some other number such as 2, 3, or 4?

13. Name a variable other than those mentioned in this chapter that you think would produce data skewed to the *right* if drawn as a polygon. Explain why.

14. Name a variable other than those mentioned in this chapter that you think would produce data skewed to the *left* if drawn as a polygon. Explain why.

15. Name a variable other than those mentioned in this chapter that you think would produce data that would be normally distributed. Explain why.

Chapter 4

The Mean:
The Most Popular Average

In this chapter, you will learn how to compute the most popular average, what it really tells us, and when it should and should not be used.

The most widely used average is the *arithmetic mean*, usually called the *mean*. Computing it is simple: just sum (add up) the scores and divide by the number of scores. This is the formula:

$$M = \frac{\sum X}{N}$$

Where:
M = mean
X = scores
N = number of scores

Note that the Greek letter Σ (sigma) is pronounced "sum of" in statistics. Although most applied researchers use M as the symbol for the mean, many statisticians use this symbol, which is pronounced "X-bar":

$$\bar{X}$$

Although computing the mean is easy, understanding its meaning is a bit harder. Since its meaning has important implications, we'll consider it carefully. The mean is defined as "the point around which the deviation scores sum to zero." Now that's a mouthful! Let's see what it means with an example. Let's suppose these scores are the number of cents donated by kindergarten children to charity:

4, 8, 8, 9, 11, 14, 16

For these scores, the mean is 10.0. (Since the scores sum to 70 and there are 7 scores, 70/7 = 10.0). Now, if we subtract the mean from each score, we have what are called the *deviation scores*, which indicate the number of points that each score differs from the mean. The symbol for the deviation scores is a lower case x, to differentiate it from the scores themselves, which are represented by an upper case X. The scores we are considering and the corresponding deviation scores are shown in Table 4.1.

Table 4.1 Scores and Deviation Scores

Score (X)	Minus	Mean	Equals	Deviation Score (x)
16	–	10.0	=	6.0
14	–	10.0	=	4.0
11	–	10.0	=	1.0
9	–	10.0	=	–1.0
8	–	10.0	=	–2.0
8	–	10.0	=	–2.0
4	–	10.0	=	–6.0
				Sum = 0.0

Notice that the deviation scores in the last column sum to 0.0. This result is not unique to this set of scores. Indeed, *for any set of scores*, the sum of the deviations from the mean equals zero. Thus, the mean is the value that has an equal number of *deviation points* above it (1.0 + 4.0 + 6.0 = 11.0) and below it (–6.0 +–2.0 ╷ –2.0 + –1.0 ⁻ 11.0). That is, the sum of the deviations above the mean equals the sum of the deviations below the mean. It's important to note that the mean is *not* defined as the value that has an equal number of cases above and below it. Indeed in our example, there are four cases below the mean (with scores of 4, 8, 8, and 9) and only three cases above the mean (with scores of 11, 14, and 16).[1]

Why should we care that the mean has an equal number of deviation points instead of an equal number of cases on both sides of it? Simply because one case or a small percentage of cases with very extreme scores can greatly affect the mean. Here's an example: suppose the highest score in our example was 80 cents instead of 16 cents. Then these are the scores and their mean:

$$4, 8, 8, 9, 11, 14, 80 \qquad M = 19.1$$

---

[1]In a perfectly normal, symmetrical distribution, the mean has equal numbers of cases on both sides of it. In a skewed distribution, it does not.

Notice that the mean has been pulled up by the one case with a large number of deviation points, the case with a score of 80. Is 19.1 a good or representative average? No, because none of the children gave close to 19 cents and because 6 of the 7 children gave less than 19 cents.

A small number of cases that are very different from the others, either because they are very high or very low, are called *outliers*. In the example we just considered, a score of 80 is clearly an outlier. Outliers pull the mean toward them and, if sufficiently different from the other values, may make the mean unrepresentative. This occurs because the mean must balance the number of deviation points (and not the number of cases) on each side.

You may have already guessed that outliers create a tail that produces a skewed distribution. You learned about skewed distributions in the last chapter. Thus, when a distribution is highly skewed, the mean is *not* a good choice as an average. We'll explore alternative averages that should be used for describing skewed distributions in the next chapter.

There's one other important implication derived from the fact that the mean is the balance point of the number of deviation points on each side of it: the mean is appropriate only for use with *equal interval* data, which we explored in Chapter 1. It makes no sense to talk about the balance point among the deviation points if the differences among scores represent different amounts of the variable we have measured. Thus, for *nominal* and *ordinal* data, we will need to use one of the alternative averages presented in the next chapter.

If the mean is subject to undue influence by skewness, why is it the most popular? First, because the mean is appropriate for use with the normal curve, which is widely found in research. In addition, the mean is associated with other very important statistics such as the standard deviation, which we'll examine in Chapter 6. For this reason, even when a distribution is only "approximately normal" because it has some skewness, researchers tend to treat it as normal and compute the mean as the average. Thus, for the emotional health scores of the foster care adolescent males we examined in Chapters 1 and 3, a researcher would be likely to consider the distribution as approximately normal and compute the mean. As it turns out, the mean emotional health score of the males is 43.3. How does this help us understand the emotional health of these boys? Well, it also turns out that the scale used to measure emotional health had been nationally standardized with a non-institutionalized (not foster care home) sample of adolescent males. For this norm group, the mean was 50.0. Thus, by comparing the two means, we can see that the foster care males have a lower average on emotional health than the national norm group. Notice, however, that this comparison of means tells only part of the story. If you refer to Table 3.1 on page 13, you will notice that a good percentage of the foster care males had scores as high as or higher than the national mean of 50.0. Thus, while means allow us to be very concise, they do not convey as much information as frequency distributions, polygons, or histograms. In Chapter 6, you'll learn that we usually report the mean with its first cousin, the *standard deviation*, in order to provide readers with a better understanding of a distribution than they can get by considering only its mean.

When analyzing a set of scores, the first thing naive people usually think of is to compute the mean. You now know that the first thing they should do is examine the shape of the distribution by preparing a frequency distribution and polygon. If the shape is normal or approximately normal and if the data are equal interval, it is appropriate to compute the mean. Otherwise, one of the alternative averages discussed in the next chapter should be used.

EXERCISE FOR CHAPTER 4

Factual Questions

1. To one decimal place, what is the mean of the scores in the box below?

    0, 2, 4, 6, 9, 9

2. From reading this chapter, you know that the deviations around the mean you computed in question 1 should sum to zero. Demonstrate that this is true by preparing for the scores for question 1 a table like Table 4.1 in this chapter.

3. Does a mean always have an equal number of cases on both sides of it?

4. What is the name for a score that is much higher or much lower than the other scores?

5. Is the mean a good choice as an average when a distribution is highly skewed?

6. For which type of data is the mean appropriate?
   A. nominal   B. ordinal   C. equal interval

Questions for Discussion

7. Name a variable for which the mean would be an appropriate measure of the average because the variable is equal interval and you do not think the distribution would be skewed.

8. Suppose a friend collected some equal interval data and computed a mean without first examining the shape of the distribution. How would you explain to your friend the desirability of first examining the shape?

NOTES

Chapter 5

The Median and Mode: Alternatives to the Mean

As you learned in the previous chapter, the *mean* is the balance point in a distribution—that is, it is the point at which the deviations below the mean balance (or cancel out) the deviations above the mean. Although the mean is the most popular average, there are two conditions under which it should not be used: (1) when a distribution is clearly skewed, and (2) when the data are not equal interval. In this chapter, you will learn about two alternative averages that can be used when the mean is inappropriate.

The Median

The *median* is the average that is defined as the *middle score* (that is, the midpoint) in a distribution of ranked scores. As the middle score, it is the point that has half the scores above it and half the scores below it. Here's a simple example from the previous chapter that includes an outlier score among children's contributions to a charity expressed in cents:

$$4, 8, 8, 9, 11, 14, 80$$

We saw that the *mean* contribution is 19.1, which is not very representative of the amounts donated since none of the children gave anything like 19 cents. The *median* contribution, on the other hand, is 9, which is quite representative.

To determine the median, first put the scores in order from low to high. Then:

(1) When the number of scores is odd, the median is the middle score. (Note that there are three scores below 9 and three scores above 9 in the above example.)

or

(2) When the number of scores is even, sum the middle two scores and divide by 2. (For example, for the scores 0, 4, 5, and 12, the middle

two scores are 4 and 5. Summing 4 and 5 and dividing by 2, we get 9/2 = 4.5, which is the median.)

A minor complication arises when there are ties in the middle (that is, when two or more cases have the same score in the area where the median lies). Here's an example:

3, 6, 7, 7, 7, 20

As you can see, when you count to the middle (three scores up or three scores down), you come to 7. Two of the 7s are in the middle of the distribution, but one of them is "above" the middle. What is the median? Well, we can use the rule we used earlier to get an approximation. Since there is an even number of scores and the middle two scores are 7, we sum them and divide by 2: 7 + 7 = 14/2 = 7, which is the approximate median. For all practical purposes, this approximation is usually more than adequate. (A method for taking the ties into account is presented in Appendix A. If you apply the method in Appendix A to the scores in this example, you will get a median of 6.8, which is very close to the value of 7 we obtained using a much easier method.)

What if there is an odd number of scores with a tie in the middle? For an approximation, we can apply the rule described on page 27. Here's an example:

1, 5, 6, 7, 8, 8, 8, 11, 50

Since there are 9 scores, we count up 5 or down 5 and come to 8, which is our approximate median. (Using the more difficult method in Appendix A, we would get 7.7, which rounds to 8, illustrating again that our approximation method is quite sound.)

When should you use the median? Under two circumstances:

(1) when analyzing equal interval data for which the mean is not appropriate because the distribution is highly skewed, and

(2) when analyzing ordinal data. As you recall from Chapter 1, ordinal data put cases in rank order, such as teachers' rankings of a list of ten discipline problems from 1 (most important) to 10 (least important). For each type of problem, such as "hitting another child," we could calculate the median rank. The medians would allow us to report which problem the average teacher thought was the most important, which one was the next most important, and so on.

The Mode

The last average we will consider is the *mode*. It is defined as the *most frequently occurring score*. Here's an example we looked at earlier in this chapter, where we found that the median is 8:

$$1, 5, 6, 7, 8, 8, 8, 11, 50$$

The mode is also 8 because 8 occurs more often than any other score. Note that the mode does not always have the same value as the median and, thus, does not always have an equal number of cases on each side of it. Here's another example we looked at earlier in this chapter, where we found that the median is 9:

$$4, 8, 8, 9, 11, 14, 80$$

For this example, the mode is 8, which does not have an equal number of cases on both sides of it.

A strength of the mode is that it is easy to determine. However, the mode has several serious weaknesses. First, a distribution may have more than one mode. Here's an example, where the modes are 9 and 10:

$$6, 6, 8, 9, 9, 9, 10, 10, 10, 15$$

Since we want a single average, the mode is not a good choice here. Second, for a small population, each score may occur only once, in which case, all scores are the mode since all occur equally often. For these reasons, the mode is seldom used. Instead, almost all researchers use either the mean or median as the average.

Concluding Comment

The mean, median, and mode all belong to the "family" of statistics called "averages." They constitute a family because they all are designed to present *one type* of information. A more formal name for this family is *measures of central tendency*. In the next chapter, we will begin our consideration of another family of statistics, *measures of variability*.

EXERCISE FOR CHAPTER 5

Factual Questions

1. Which average is defined as the *middle score*?

2. Suppose you read that the median price of a house in Mudsville is $167,000. What percentage of the houses in Mudsville cost more than $167,000?

3. What is the median of the scores shown in the box immediately below?

    12, 14, 14, 15, 20, 22, 66

4. What is the median of the scores shown in the box immediately below?

    9, 7, 10, 4, 8, 6

5. When analyzing equal interval data that is highly skewed, which average should you use?

6. When analyzing ordinal data, which average should you use?

7. How is the mode defined?

8. What is the mode of the scores for question 3?

9. For a given distribution, is it possible to have more than one mode?

10. What is another name for the family of statistics called *averages*?

Questions for Discussion

11. In question 3 you computed a median. Compute the mean for the same scores. Which average is more representative? Why?

12. Suppose you asked clinical psychologists to rank the problems they treat in their order of difficulty. That is, you asked them to give a rank of 1 to the most difficult problem, a rank of 2 to the next most difficult problem, and so on. Which average would you use when analyzing the data? Why?

13. Suppose a counselor told you that "the average salary for a beginning professional in your chosen field is $34,000." Would you be interested in knowing whether the average is the mean, median, or mode? Why? Why not?

NOTES

Chapter 6

The Standard Deviation: First Cousin of the Mean

As you learned in the previous chapter, the family of statistics called *averages* indicates where the center of a distribution lies. For the distribution of emotional health scores for a group of adolescent males in a foster care home (see page 13 of this book), the mean score is 43.3. For the national norm group on which the emotional health scale was standardized, the mean is 50.0. Thus, we know that the group in the foster care home is lower *on the average* than the national norm group. However, suppose that this is all that we know. That is, suppose we read a report and were given only the means without the frequency distribution, polygon, histogram, or any other statistics. We would be missing a lot of important information, wouldn't we? For example, the distribution on page 13 shows that a substantial percentage of the foster home boys had scores as high as or higher than the national mean of 50. Thus, clearly if we are going to report the mean (without a frequency distribution, polygon, or histogram), we'll need to supplement it with an additional statistic that indicates how spread out the scores are around their mean of 43.3. In this chapter, we will examine the statistic that is most widely used for this purpose.

The Concept of Variability

Variability refers to the differences among a set of scores. For example, if we say that there is *no variability* among a set of scores, we are saying that all the scores are the same (that is, there are no differences). Synonyms for the term *variability* are *spread* and *dispersion*.

One way we could describe the variability of the scores on page 13 would be to say that the scores range from the 3–7 score interval to the 73–77 score interval. While this helps us get a mental picture of the distribution, it does not tell us as much as the statistic we will consider in this chapter—the *standard deviation.*

How Does the Standard Deviation Measure Variability?

The standard deviation is a measure of the variability of the scores in relation to the mean of the group. In other words, the larger the standard deviation, the greater the differences between the individual scores and the mean of the scores. Consider the following two sets of scores.

SCORES FOR GROUP A: 15, 20, 25, 30, 35

SCORES FOR GROUP B: 23, 24, 25, 26, 27

Notice that the two groups have the same mean, which is 25.0. Thus, on the average, the two groups are the same, but Group A's scores are more variable than Group B's scores, which are clustered tightly around the mean of 25. Thus, the standard deviation of Group A's scores will be larger than the standard deviation of Group B's scores, as we will see next.

Computation of the Standard Deviation

To compute the standard deviation of Group A's scores, we need to set up a worktable such as that shown in Table 6.1, where the first column lists the scores and the next three columns are used to get the deviation scores, as we saw in our discussion of the mean in Chapter 4. The symbol for the deviaiton scores is the lower case x. The last column contains the squared deviation scores. We need the sum of this column in the formula for the standard deviation.

Table 6.1 Worktable for Computing the Standard Deviation of Group A's Scores

Scores (X)	Minus	Mean	Equals	Deviation scores (x)	Squared deviation scores (x^2)
35	−	25.0	=	10.0	100.0
30	−	25.0	=	5.0	25.0
25	−	25.0	=	0.0	0.0
20	−	25.0	=	−5.0	25.0
15	−	25.0	=	−10.0	100.0
					Sum = 250.0

The formula for the standard deviation is:

$$SD = \sqrt{\frac{\Sigma x^2}{N}}$$

The formula says that the standard deviation equals the square root of the sum of the squared deviation scores divided by the number of cases. Thus, the standard deviation for Group A equals:

$$SD = \sqrt{\frac{250}{5}} = \sqrt{50} = 7.07 = 7.1$$

Now let's compute the standard deviation for Group B, which we already decided would be smaller because the scores have less variation.

Table 6.2 Worktable for Computing the Standard Deviation of Group B's Scores

Scores (X)	Minus	Mean	Equals	Deviation scores (x)	Squared deviation scores (x^2)
27	–	25.0	=	2.0	4.0
26	–	25.0	=	1.0	1.0
25	–	25.0	=	0.0	0.0
24	–	25.0	=	–1.0	1.0
23	–	25.0	=	–2.0	4.0
					Sum = 10.0

Substituting the sum from Table 6.2 and the number of cases into the formula, we get:

$$SD = \sqrt{\frac{\Sigma x^2}{N}} = \sqrt{\frac{10.0}{5}} = \sqrt{2.0} = 1.41 = 1.4$$

Just as we expected, Group B's standard deviation (1.4) is considerably smaller than Group A's standard deviation (7.1). Thus, if we were reading a report on this data, we would find the statistics in Table 6.3. Based on

Table 6.3 Means and Standard Deviations for Groups A and B

Group	M	SD
A (N = 5)	25.0	7.1
B (N = 5)	25.0	1.4

this information, we would know that the two groups are the same on the average but that the scores of Group A are considerably more variable (spread out or dispersed) than those of Group B.

Note that there are several formulas for the standard deviation. The one that we just examined most clearly illustrates what makes the standard deviation large or small (that is, differences from the mean). An algebraically equivalent formula that is more convenient for computations is presented in Appendix B. When computing the standard deviation for a large number of scores, you'll find it easier to use the formula in Appendix B.[1]

The Standard Deviation and the Normal Curve

You were introduced to the normal curve in Chapter 3. It's the bell-shaped curve that abounds in nature. For example, the heights of ten-year-old girls, and the weights of Idaho potatoes are normally distributed. The annual rainfall in the United States over the years and the annual amount of snow in Siberia over the years are normally distributed. In addition, many social and behavioral scientists believe that cognitive traits such as achievement and aptitude as well as traits of temperament such as depression and cheerfulness are probably normally distributed. In fact, most standardized measures of these traits are designed to yield normal distributions in national norm groups.[2]

It turns out that the standard deviation has a special relationship to the normal curve. Because of this relationship, if we know the mean and standard deviation of a normal distribution, we know a great deal about the distribution. Let's take a careful look at this relationship.

The standard deviation is a *unit* that we can use for thinking about the normal curve. For example, suppose that for a normal distribution the mean equals 50.0 and the standard deviation equals 10.0. (We would determine these values using the formulas in this book.) Because the standard deviation equals 10.0, *one standard deviation unit* for this distribution equals 10 points, *two standard deviation units* equal 20 points, and *three standard deviation units* equal 30 points. Figure 6.1 shows that

---

[1] A slightly different version of the formula for the standard deviation should be used when computing the standard deviation for a sample. This formula is not presented in this book because it is for use with inferential statistics, which are not covered in this book.

[2] This is done, in part, by pretesting potential items for a measure and then selecting those items that will yield a normal distribution of scores. In addition, scores are adjusted during norming to create normalized distributions.

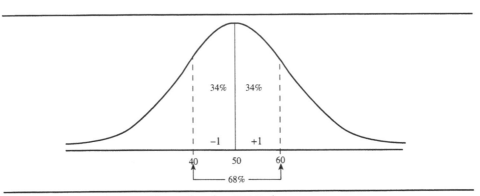

Figure 6.1 Normal Curve Illustrating the 68% Rule

if you go out one standard deviation unit on both sides of the mean, you collect about 68% of the cases in the normal distribution.

Let's pause to consider this again since it's sometimes hard to grasp at first. Let's say you have a normal distribution (the bell-shaped polygon) and you use the formulas you learned in this book to calculate the mean and standard deviation. Whatever the values of the mean and standard deviation you obtain, if you add the standard deviation to the mean and then subtract the standard deviation from the mean, you will identify the middle area with 68% of the cases. This is always true for a normal distribution (a bit like magic). Put in the reverse, if you *don't* have 68% of the cases when you do this, you *don't* have a normal distribution. Thus, this is an invariant characteristic of the normal distribution. We call this the "*68% rule*" or the "two-thirds rule-of-thumb" since 68% is just a little over two-thirds.

How does knowing the 68% rule help us? By letting us picture the spread of the scores after being given only two statistics. For example, you might read in a report only this: "$M = 50.0$, $SD = 10.0$, the distribution is normal." We could then picture that 50.0 is the score in the middle of the distribution and that 68% of the cases lie between 40.0 (50.0 − 10.0 = 40.0) and 60.0 (50.0 + 10.0 = 60.0). The remaining 32% of the cases are, of course, spread out even farther below 40.0 and above 60.0.

Another rule that helps us picture a normal distribution of scores is this: if you go out two standard deviation units, you gather up approximately the middle 95% of the cases.[3] In our example, one standard

---

[3]For precisely 95%, we go out 1.96 standard deviation units instead of 2 units.

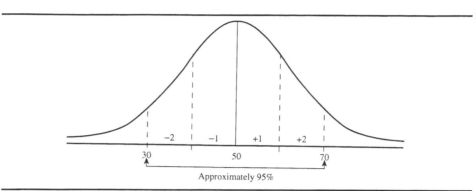

Figure 6.2 Normal Curve Illustrating the 95% Rule

deviation unit equals 10.0; thus, two units equals 20. Figure 6.2 illustrates the *95% rule*.

Figure 6.3 illustrates the *99.7% rule*. If you go out three standard deviation units (in our example, 30 points) on both sides of the mean, you gather up the middle 99.7% of the cases.

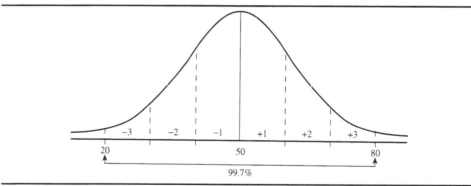

Figure 6.3 Normal Curve Illustrating the 99.7% Rule

In a normal distribution, these three rules hold regardless of the value of the mean and standard deviation that you compute. They are important for helping us picture a distribution, so let's look at one more example. Suppose you had a normal distribution and calculated the mean to be 100.0 and the standard deviation to be 20.0. Then you would know that 68% of the cases lie between 80.0 and 120.0 (that is, 100.0 +/– 20.0). You would also know that 95% of the cases lie between 60.0 and 140.0 [that is, 100 +/– (2)(20)]. Finally, you would know that 99.7% of the cases lie

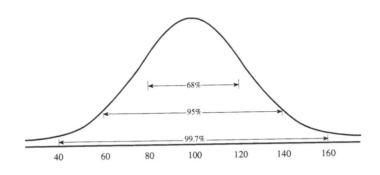

Figure 6.4 Normal Curve Illustrating Three Rules for a Normal Distribution with a Mean of 100 and a Standard Deviation of 20

between 40.0 and 160.0 [that is, 100 +/– (3)(20)]. These scores and percentages are shown in Figure 6.4.

Concluding Comment

What about the emotional health scores we started with on page 1 of this book, which gave us the distribution on page 13? Well, we already know that the mean of the distribution is 43.3, and the mean of the national norm group is 50.0. It turns out that the standard deviation of the foster care group equals 16.2, while the standard deviation of the national norm group is 15.0. This information is summarized in Table 6.2.

Table 6.2 Means and Standard Deviations of Emotional Health Scores

Group	M	SD
Foster care males	43.3	16.2
Norm group males	50.0	15.0

Now we know that the foster care males are more variable than the national sample. In addition, we can see that there is considerable overlap between the two sets of scores. We can see this by using the 68% rule and applying it to the norm group: 50.0 – 15.0 = 35.0, and 50.0 + 15.0 = 65.0, which tells us that about 68% of the norm group have scores between 35.0 and 65.0—an interval that includes the foster care males' mean of 43.3.

We can also apply the 68% rule to the values for the foster care males: 43.3 − 16.2 = 27.1, and 43.3 + 16.2 = 59.5. We see that the middle 68% lies between 27.1 and 59.5—indicating that many of the foster care males had scores higher than the norm group mean of 50.0.

Of course, there's a limitation to what we've just done here. Remember that we decided in Chapter 3 that the distribution of emotional health scores of the foster care males is not perfectly normal but only "approximately normal." Thus, the application of the 68% rule is only a *very rough* rule of thumb, yet it is useful when thinking about the variability in a distribution and when comparing two distributions.

The standard deviation should be used as a measure of variability only when the mean has been selected as the average—and the mean should be used only with equal interval data whose distribution is not severely skewed. In the next chapter, we will consider other measures of variability that can be used when the standard deviation is inappropriate.

EXERCISE FOR CHAPTER 6

Factual Questions

1. In this chapter, what two synonyms are given for the term *variability*?

2. Which group shown in the box immediately below has more variability among its scores? (You should be able to determine the answer through inspection, without calculating the standard deviation.)

> Scores for Group X: 89, 91, 87, 94, 92, 88, 87, 90
> Scores for Group Y: 50, 58, 42, 40, 38, 60, 44, 32

3. If all members of a group have the same score, what is the value of the standard deviation of their scores?

4. What is the value of the standard deviation for the scores shown in the box immediately below?

> Scores for a group: 5, 4, 6, 8, 2

5. According to the information in Table 6.3, are the scores for Group C or the scores for Group D more variable?

Table 6.3 Statistics for Question 5

Group	N	M	SD
C	33	20.1	3.4
D	43	25.2	2.7

6. If $M = 52.0$ and $SD = 4.0$ for a normal distribution, between which two values do 68% of the cases lie?

7. If you go out two standard deviation units on both sides of the mean, you will gather up what percentage of the cases?

8. If $M = 133.9$ and $SD = 20.5$, what percentage of the cases in a normal distribution lies between scores of 113.4 and 154.4?

9. If $M = 30.0$ and $SD = 6.0$ for a normal distribution, between which two values do 99.7% of the cases lie?

10. If you go out one standard deviation unit on both sides of the mean, you will gather what percentage of the cases?

Question for Discussion

11. Name a variable you might study for which you think the standard deviation would be the appropriate measure of variability to compute. Explain your choice.

NOTES

Chapter 7

The Range, Quartiles, and Interquartile Range

You learned in the previous chapter that the standard deviation is a measure of variability based on the differences between the individual scores and the mean of the scores. In a research report, the mean and standard deviation are usually reported together to describe the average and the variability, respectively.

Clearly, if it is inappropriate to compute the mean (because a distribution is highly skewed or the data are not equal interval), it is also inappropriate to compute the standard deviation. In this chapter, you will learn about alternative measures of variability to use when the standard deviation is inappropriate.

The Range

The roughest measure of variability is the *range*; it is defined as the number of score points covered in a distribution. The range is easy to compute: subtract the lowest score from the highest score and add 1. Consider these scores:

$$2, 4, 5, 6$$

The range is $6 - 2 = 4 + 1 = 5$.

Some students wonder why we add 1. The answer can best be seen by example. From the score of 2 through the score of 6 in our example, there are 5 score points. The points are 2, 3, 4, 5, and 6 (a total of 5 numbers). Hence, the range is 5, which was obtained by adding 1 to the difference between the highest and lowest scores.

Obviously, the range is based on only two scores — the two most extreme scores. This is a weakness, especially when a distribution is skewed or when there are outliers. This is illustrated by considering the following scores, which are the numbers of days that elapsed from the time some

43

women noticed a breast cancer symptom to the time they made an appointment with a physician.

1, 2, 3, 3, 4, 4, 14, 17, 27, 28, 33, 50, 128, 130, 244, 260

The scores range from 1 to 260 days, yielding a range of 260 days.[1]

It's a little easier to see what's going on here if we convert each of the scores from days of waiting to months of waiting by dividing each score by 30 (the average number of days per month) and rounding to one decimal place. Doing this, we get these months of waiting:[2]

.0, .1, .1, .1, .1, .1, .5, .6, .9, .9, 1.1, 1.7, 4.3, 4.3, 8.1, 8.7

This tells us that the number of months ranged from 0 to 8.7, for a range of 9.7 months. Yet, the majority of women called within one month, and only two waited more than 4.3 months. Clearly, the most extreme score (8.7, which is one of the outliers) has had an undue influence on the range (9.7). Thus, the range indicates that the group as a whole is much more variable than it really is.

As you can see, the distribution of months of waiting is skewed. This suggests a principle: the range is not a good measure of variability when a distribution is skewed. In addition, even if a distribution is *not* skewed the range is still not a good choice because measurement theory tells us that the more extreme the score, the more likely it is in error. That is, we can get extreme scores for two reasons: (1) some people are truly very different from the others, and (2) some measurement errors are very large. For example, large errors — such as a student not understanding the directions on a test and marking all the answers in the wrong spaces or a clerk accidentally adding a zero to a number when keyboarding data and typing 260 instead of 26 — produce extreme scores. Clearly, when describing a group, we do not want to use a statistic that is based on the two scores most likely to have been influenced by large errors (that is, the two least reliable scores). Thus, the range is seldom used by researchers except in informal communications.

---

[1]Data supplied by Dr. Diane Lauver, School of Nursing, University of Wisconsin–Madison. For instructional purposes, only a select portion of the data are presented here. For more information on this topic, see Lauver, D., & Tak, Y. (1995). Optimism and coping with a breast cancer symptom. *Nursing Research, 44*, 202-207.
[2]For 1 day, 1/30 = 0.03 = 0.0 months.

Quartiles

We can provide a better picture of the variability in a distribution by reporting *quartiles*. Quartiles are the score values that divide a distribution into quarters. To find the quartiles, first put the scores in order from low to high. Then determine how many scores are in each quarter by dividing the number of scores by 4; in the previous example of days waiting, there are 16 scores, so 16/4 = 4. To find the *first quartile* (Q_1), count up 4 scores, as shown here:

$$1, 2, 3, 3, 4, 4, 14, 17, 27, 28, 33, 50, 128, 130, 244, 260$$
$$\Uparrow$$
$$3.5$$

The first quartile is halfway between 3 and 4, which is 3.5.

To find the *second quartile* (Q_2), count up another four scores, as shown here:

$$1, 2, 3, 3, 4, 4, 14, 17, 27, 28, 33, 50, 128, 130, 244, 260$$
$$\Uparrow$$
$$22$$

The second quartile is halfway between 17 and 27, which is 22 (that is, 17 + 27 = 44/2 = 22). You may have noticed that the second quartile is a friend that we met in Chapter 5; it is the *median* (the value that has 50% of the scores above it and 50% below it). Because the distribution we are considering is skewed due to the outliers, the median is an appropriate average to report for these data.

To find the *third quartile* (Q_3), count up another four scores, as shown here:

$$1, 2, 3, 3, 4, 4, 14, 17, 27, 28, 33, 50, 128, 130, 244, 260$$
$$\Uparrow$$
$$89$$

The third quartile is halfway between 50 and 128, which is 89 (that is, 50 + 128 = 178/2 = 89).

With these calculations, we can now describe the distribution as follows:

The median is 22, $Q_1 = 3.5$, and $Q_3 = 89$.

From this, readers of a report on these data would know that the average number of days is 22. Because the average is the median, they would know that half the cases were above 22 and half were below 22. They would get a good picture of the variability of the distribution from the quartiles, which indicate that 25% of the women waited less than 3.5 days and 25% of the women waited more than 89 days.

We can provide a more complete picture by including the highest and lowest value. This gives us a *five-point description* of the distribution, in which 3.5 has been rounded to a whole number (4) for the sake of consistency with the other values, which are whole numbers:

1	4	22	89	260
Low score	Q_1	Q_2	Q_3	High score
25%	25%	(Median) 25%	25%	

With just five values (1, 4, 22, 89, and 260), we have provided a detailed summary of the distribution.

At this point, you may wonder why we don't just include all 16 scores in the research report. One reason is that researchers are obligated to discuss the distribution and its meaning when writing reports on their research, and summary statistics such as the median help readers understand the main thrust of the data. Second, we usually have more than 16 scores. In fact, the scores in this example are only 16 of the 140 scores that the researchers actually collected. (For instructional purposes, we only examined a small number of them here.) In addition, the researchers measured a number of other variables, so they had many hundreds of scores, all of which needed to be summarized with descriptive statistics.

The Interquartile Range

Instead of providing a five-point description of skewed distributions, many researchers report only the median as the average and a single statistic, the *interquartile range*, as an indicator of variability. To compute the interquartile range, first identify the first and third quartiles, which were found above in our earlier example and are shown again here:

1, 2, 3, 3, 4, 4, 14, 17, 27, 28, 33, 50, 128, 130, 244, 260
⇑ 3.5   ⇑ 89

Then subtract Q_1 (3.5) from Q_3 (89) and add 1, which yields 86.5, which is the interquartile range. As you can see, the interquartile range is a special type of range — it is the *range of the middle 50% of the scores*. In other words, in our example, the number of score points covered by the middle 50% of the scores is 86.5 days. An advantage of the interquartile range (*IQR*) over the range is that the *IQR* is *not* based on the two most extreme and unreliable scores; it is based on the more reliable middle part of the distribution.

As you can see, reporting only the median and the interquartile range is less informative than giving a five-point description. Nevertheless, researchers often report only the median and *IQR* in the interests of brevity.

A Brief Review of Chapters 1 through 7

Just in case you're getting a little lost in the details, let's review the big picture. First, when we have nominal data (that is, naming data such as people naming the candidate they plan to vote for), we usually report frequencies (numbers of cases) and percentages for each category named. When we have scores, we first prepare a frequency distribution and polygon or histogram, which may be shown in a research report if space permits. If the data are equal interval and the distribution is not seriously skewed, we usually compute the mean and standard deviation as the measures of average and variability, respectively. If the data are ordinal or if the data are seriously skewed, we usually compute the median and the interquartile range as the measures of average and variability. An alternative for ordinal or skewed data is to report a five-point description, which includes the median, the first and third quartiles, and the lowest and highest scores.

Up to this point in the book, we have been considering how to describe a group as a *whole* by organizing and summarizing the data we've collected. In the next chapter, you'll learn how to use the mean and standard deviation to interpret the scores of *individuals*.

EXERCISE FOR CHAPTER 7

Factual Questions

1. What does measurement theory tell us about extreme scores?

2. What is the value of the range for the scores shown in the box below?

8, 9, 12, 7, 10, 6, 11, 31

3. What is the value of the first quartile for the scores in question 2? (Hint: Remember to order the scores from low to high before calculating the value.)

4. What is the value of the second quartile for the scores in question 2?

5. What is the value of the third quartile for the scores in question 2?

6. What is the value of the interquartile range for the scores in question 2?

7. What is another name for the second quartile?

8. Researchers usually report the median and *IQR* for which type of distribution? (Circle one or more.)
   A. normal distributions   B. skewed distributions

9. What five points are reported in a five-point description?

Question for Discussion

10. In the question 6, you were asked to compute the interquartile range for the data in question 2. Is the interquartile range a better choice than the standard deviation for these data? Why? Why not?

Chapter 8

Standard Scores

A *raw score* is the number of points accumulated or earned. When we tell an examinee that she earned 63 raw score points on a 90-item algebra test, she is likely to ask us, "What does that mean?" One way to interpret her score is to find out how this score compares with the scores earned by a group of examinees who took the same test. A group that is used for the interpretation of scores is called a *norm group*. For most standardized tests and personality scales, the norm group is a representative national group of examinees at a certain age or grade level. However, we can also develop local norms, which allow us to compare an individual's performance with that of a local group such as all fifth graders in a school district.

In this chapter, you will learn how we can use the mean and standard deviation to develop *standard scores*, which allow us to interpret the scores of individuals in light of the performance of a norm group.

Standard Scores

Standard scores (also called z-scores) tell us how far examinees are from the mean of a norm group in standard deviation units. Sounds complicated? It's not, really, if you have already mastered the standard deviation. Figure 8.1 shows a normal distribution with a mean raw score of 60.0 and a standard deviation of 8.0. Above the raw scores are corresponding z-scores. (Note that the curve is marked off in units of 8 because one standard deviation unit equals 8 points.) The figure shows that a person with a raw score of 60 (the same score as the mean) has a z-score of 0.0. This results from the fact that the person is zero standard deviation units from the mean. A person with a raw score of 68 has a z-score of +1 because that person is one standard deviation unit (for this distribution, 8 points) above the mean. A person with a raw score of 52 has a z-score of −1.0 because he or she is one standard deviation unit below the mean.

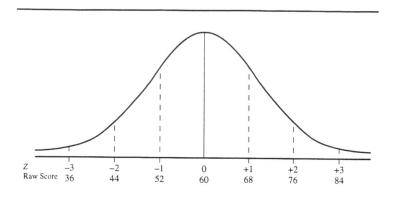

Figure 8.1 z-Scores for a Normal Distribution with a Raw Score
Mean of 60 and Standard Deviation of 8.

But what about someone with a raw score of 63 or a raw score of 48? We can determine their z-scores with this formula:

$$z = \frac{X-M}{SD}$$

Where:

X = raw score

M = mean

SD = standard deviation

If you think about the formula, it makes sense. The numerator determines the number of raw score points an examinee is from the mean. By dividing this by the standard deviation, we determine the number of standard deviation units the examinee is from the mean.

Here's how the formula works for an examinee with a raw score of 63:

$$z = \frac{63-60.0}{8.0} = \frac{3}{8} = .38$$

This tells us that the examinee is .38 of a standard deviation unit above the mean. Note that we usually calculate z-scores to two decimal places.

Here's how the formula works for an examinee with a raw score of 48:

$$z = \frac{48-60.0}{8.0} = \frac{-12}{8.0} = -1.50$$

This tells us that the examinee is 1.50 standard deviation units *below* the mean; hence, the z-score is negative (−1.50).

Transformed Standard Scores

Of course, it's not nice to report a z-score of zero to an average examinee who has worked hard on a test. Worse, it's unacceptable to report a negative z-score to an examinee who is below the mean but got many items right on a test. Thus, we almost always transform z-scores to a transformed (modified) standard-score scale for which the average is above zero and there are no negatives. We'll call these *transformed standard scores* (*TSS*) in this book.

Here's how the College Board transforms the z-scores before reporting them: they multiply each person's z-score by 100 and add 500 to the sum. For an examinee with a z-score of .33, her College Entrance Examination Board (CEEB) score is:

$$.38\,(100) + 500 = 538$$

The formula for *CEEB scores* is:

$$TSS_{CEEB} = (z)(100) + 500$$

Applying the formula to a z-score of -1.50, we get:

$$TSS_{CEEB} = (-1.50)(100) + 500 = 350$$

Figure 8.2 shows CEEB scores in relation to z-scores.

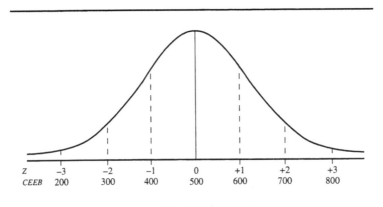

z	-3	-2	-1	0	$+1$	$+2$	$+3$
CEEB	200	300	400	500	600	700	800

Figure 8.2 CEEB Scores and z-Scores in a Normal Distribution

As you can see in Figure 8.2, the mean CEEB score is 500 and the standard deviation is 100 points. This is always true of CEEB scores because the College Board always computes z-scores for all examinees who take one of their tests, and then multiplies each z-score by 100 and adds

500 to each product. This leads to two principles: the constant by which you multiply all the z-scores becomes the new standard deviation; the constant that you add to the products becomes the new mean.

The general formula for transformed standard scores is:

TSS = z-score times the new standard deviation plus the new mean

In symbols, this becomes:

$$TSS = (z)(SD_{new}) + (M_{new})$$

An important advantage of transformed standard scores such as CEEB scores is that they let us make legitimate comparisons *across tests*. For example, two of the College Board tests are *Verbal* and *Mathematical*. Since the CEEB scores for both tests are based on a single norm group, we can determine areas of strength and weakness by comparing an examinee's two scores. For example, if Sean has a 621 on the Verbal and a 455 on the Mathematical, it is appropriate to conclude that relative to the norm group, he is stronger in the verbal area than in the mathematical area.

When building a test, you can select any new standard deviation and new mean you desire. The College Board selected 100 and 500. Makers of intelligence tests (IQ tests) usually use 15 and 100. Here's what they do to get the standard scores: First they administer their new test to a national group (the norm group) and get the raw scores. Next they convert each raw score to a z-score. Then they multiply each z-score by 15 and add 100 to the product. Thus, a z-score of .38 becomes an *IQ score* of 106 [(.38)(15) + 100 = 105.7 = 106]. (Note that we usually round transformed standard scores to whole numbers.) The formula for IQ scores is:

$$TSS_{IQ} = (z)(15) + 100$$

Applying the formula to a z-score of -1.50, we get:

$$TSS_{IQ} = (-1.50)(15) + 100 = 77.5 = 78$$

Since we have multiplied the z-scores by 15, the standard deviation of IQ scores in the norm group equals 15, and since we added 100, the mean of IQ scores is 100. This is shown in Figure 8.3.

For consumers who will be using an IQ test, the test makers prepare a norms table that shows each possible raw score with its corresponding IQ score that was calculated for those in the norm group. Thus, when administering an IQ test to a client, all a psychologist has to do is count the number of raw score points earned, go to the norms table in the test manual and read the corresponding IQ score. By knowing the mean (100) and standard deviation (15) of IQ scores, the IQ score for the client is easy to interpret. Considering Figure 8.3, we can see that an IQ of 106 is only very slightly above average and that an IQ of 78 is substantially (more than one standard deviation) below average.

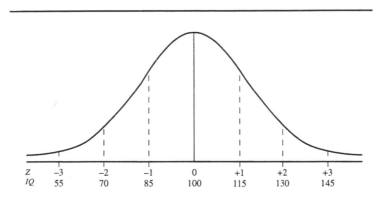

| z | −3 | −2 | −1 | 0 | +1 | +2 | +3 |
| IQ | 55 | 70 | 85 | 100 | 115 | 130 | 145 |

Figure 8.3   IQ Scores and z-Scores in a Normal Distribution

A general purpose standard score that is used for many standardized measures is one that has a mean of 50 and a standard deviation of 10. These values were suggested by a statistician named McCall, who referred to the resulting transformed standard scores as *T* scores. Thus, they are often called *McCall's T scores*. (In test manuals, they are often simply called *standard scores*.) The formula for McCall's *T* scores is:

$$TSS_{\text{McCall's } T} = (z)(10) + 50$$

Applying the formula to a z-score of −1.50, we get:

$$TSS_{\text{McCall's } T} = (−1.50)(10) + 50 = 35$$

Figure 8.4 shows McCall's *T* scores in relation to z-scores.

Now let's consider again the emotional health scores for the foster care adolescent males presented on the first page of this book. It turns out that the scores you see there are transformed standard scores. The developer of the test developed norms by converting all of the raw scores in a national norm group to z-scores, multiplying each z-score by 15 and adding 50. Thus, in the norm group, the emotional health scores were distributed as shown in Figure 8.5. Using the norms table for the test, the researchers

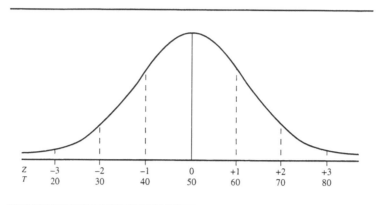

Figure 8.4 McCall's *T* Scores and *z*-Scores in a Normal

were able to obtain the foster care males' transformed standard scores, which are reported on page 1 of this book.

We can use the information in Figure 8.5 to interpret the scores on page 1. For example, the first score on page 1 is 63. Referring to Figure 8.5, we can see that this score is close to one standard deviation above the mean. Thus, we could describe this score as "high average" or "very high average," if we treat the area from a *z*-score of -1.0 to 1.0 as the "average area." The second score on page 1 is 28. We can see that in Figure 8.5 this score is more than one standard deviation below the mean, which may be interpreted as being substantially below average.

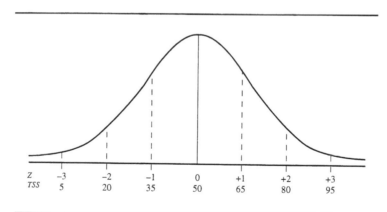

Figure 8.5 *TSS* Scores for Emotional Health and *z*-Scores in a
National Norm Group

In this chapter, you have seen how useful standard scores are for interpreting the performance of individuals.

EXERCISE FOR CHAPTER 8

Factual Questions

1. If Joan has a raw score that is identical to the mean of the norm group, what is her *z*-score?

2. If Jake has a *z*-score of 2.0, his score is how many standard deviation units from the mean?

3. If Jennifer has a *z*-score of –1.0, is she above or below the mean?

4. If the raw score mean and standard deviation of a group are 20.0 and 2.3, respectively, what is the *z*-score of a person with a raw score of 16?

5. If Julie has a *z*-score of 2.50, what is her CEEB score?

6. If Justin has a *z*-score of –.50, what is his IQ score?

7. If Jessica has a *z*-score of 1.25, what is her *T* score?

Question for Discussion

8. The last emotional health score on page 1 of this book is 34. Explain how you would interpret this score. (Keep in mind that the scores are McCall's *T* scores that are based on the performance of a national norm group.)

NOTES

Chapter 9

Scattergrams and the Pearson *r*

In this chapter, we will consider how to describe the relationship between two sets of scores. A *variable* is a trait or characteristic on which individuals differ or *vary*. For example, height is a variable—people vary in height because some are taller than others. Weight is another variable. Are height and weight related? Yes. Generally, taller people weigh more and shorter people weigh less. Thus, we can say that there is a direct (or positive) relationship between the two variables, but, as you know, there will be many exceptions. For example, some short people weigh more than some tall people. The number and size of the exceptions need to be taken into account when describing the relationship between two variables. We'll look at two methods for describing relationships in this chapter.

Scattergrams

A *scattergram* (also known as a *scatter diagram* or *scatterplot*) is a figure that shows the relationship between two variables. Let's start with a simple example for the height and weight of eight people in Table 9.1.

Table 9.1 Height and Weight for Eight People

Person	Height (inches)	Weight (pounds)
Jose	72	199
Bill	69	185
Millie	66	190
Kelly	63	155
Frank	62	140
Diana	60	128
Willy	57	132
Jean	54	117

As you can see in Table 9.1, taller people such as Jose and Bill weigh more, and shorter people such as Willy and Jean weigh less, yet there are exceptions such as Diana, who is taller than Willy but weighs less than Willy. The number and size of the exceptions can be seen on a scattergram such as the one in Figure 9.1. It was created by placing a dot for each person where his or her two scores (height score and weight score) intersect. To help you understand scattergrams, the dot for Jose has been labeled with his name; his dot is where a height of 72 inches meets a weight of 199 pounds.

Figure 9.1 indicates that there is a *direct* (positive) relationship between height and weight. The pattern from the lower-left corner to the upper-right corner indicates that as height goes up, weight also goes up. If the dots formed a straight line, the relationship would be perfect. Since there is scatter (instead of a straight line), the relationship is less than perfect. The more scatter, the weaker the relationship is.

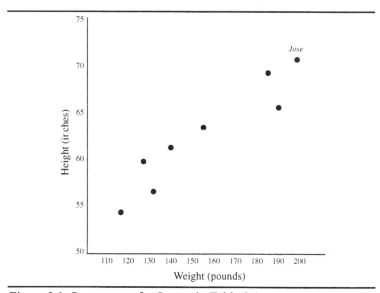

Figure 9.1 Scattergram for Scores in Table 9.1

At this point, you may not fully appreciate scattergrams because we've been considering a small number of scores, which are easy to eyeball in order to determine the relationship, so let's consider a larger group: the 58 adolescent foster care males that we started with at the beginning of this book. In addition to measuring their emotional health, the researchers also measured their social relationships with a scale that asked them to respond to statements such as, "I prefer being alone than with other kids my age" and "Being together with other people gives me a good feeling." The responses were scored in such a way that higher scores indicate better social relationships. The two scores for each male are shown in Table 9.2 on the next page. As you can see, it's hard to eyeball such a large amount of data to determine whether there's a relationship.

Figure 9.2 is a scattergram for the data in Table 9.2. It shows clearly that there is a direct relationship between emotional health and social relationship scores because the majority of the dots form a pattern that goes from the lower-left corner to the upper-right corner. However, the relationship in Figure 9.2 is much weaker than the one in Figure 9.1 because there is much more scatter in Figure 9.2. We'll come back to the data in Figure 9.2 a little later in this chapter and interpret it more precisely.

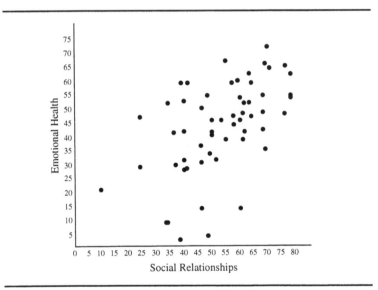

Figure 9.2 Scattergram for Scores in Table 9.2

So far, we've seen two *direct* relationships (also called *positive* relationships). In a direct relationship, those who have high scores on one variable tend to have high scores on the other *and* those who have low scores on one variable tend to have low scores on the other. This creates a pattern of dots from lower-left to upper-right on a scattergram.

Table 9.2 Emotional Health and Social Relationships Scores for 58 Males

Partici-pant number	Emotion-al health	Social Relation-ships	Partici-pant number	Emotion-al health	Social Relation-ships	Partici-pant number	Emotion-al health	Social Relation-ships
1	63	64	21	60	60	41	32	53
2	28	41	22	47	77	42	47	65
3	59	57	23	67	56	43	31	47
4	4	49	24	64	72	44	59	43
5	59	65	25	66	70	45	9	35
6	37	47	26	55	69	46	59	40
7	30	37	27	53	64	47	52	63
8	37	47	28	21	11	48	52	35
9	43	69	29	42	37	49	39	62
10	53	41	30	48	62	50	41	51
11	14	47	31	50	47	51	27	41
12	46	51	32	55	49	52	42	51
13	3	39	33	9	34	53	65	77
14	39	56	34	52	79	54	62	79
15	47	58	35	44	58	55	36	70
16	48	69	36	29	25	56	42	63
17	46	61	37	54	61	57	42	41
18	32	41	38	73	71	58	34	50
19	46	54	39	55	79			
20	14	61	40	47	25			

Of course, relationships can be *inverse* (or *negative*). For example, there is a perfect negative relationship between the amount of sound heard and the distance from the sound. That is, if you have a device emitting a sound at a constant volume, as you move away from the device you hear the sound less and less — without exception, because it never gets louder as you get farther away. Notice that as the distance *goes up*, the amount of sound heard *goes down*. The relationship between amount of sound and distance from the sound is illustrated with the scattergram in Figure 9.3, where you can see that the dots follow a straight line that goes from the upper-left corner to the lower-right corner. This is as strong as a

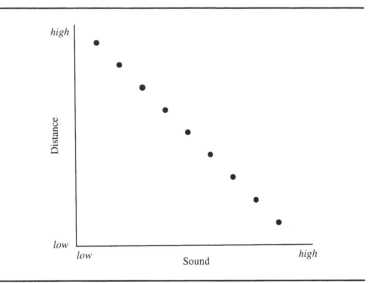

Figure 9.3 Scattergram for a Perfect, Negative Relationship

relationship can be; we call it *perfect* because there is no scatter. There is no scatter because there are no exceptions to the trend.

Of course, most relationships are not perfect. A good example is the relationship between absences from school and GPA. Many researchers have reported a negative relationship between the two variables: in general, those who are absent *more* earn *lower* GPAs, but, of course, there are many exceptions. Some students who are absent a lot are skilled and manage to make up the missed work. Others who are present every day are less skilled and, despite their perfect attendance, don't do well. Figure 9.4 shows a scattergram that is typical for this relationship. The pattern indicates an inverse relationship, but the scatter indicates that there are many exceptions; thus, we can describe it as a *weak inverse relationship*.

So far, we have seen *positive* and *negative* relationships on scattergrams. By far, these are the most common. Once in a great while, however, we observe *curvilinear* relationships, which are neither positive nor negative. Figure 9.5 illustrates a curvilinear relationship between anxiety and success on a test. At the lowest levels of anxiety, the test performance is also very low. These are people who have so little anxiety (or concern) about taking the test that they don't even try to do well. Modest levels of anxiety are associated with better test performance (enough anxiety to motivate the students to give it their best shot). Finally, high levels of anxiety

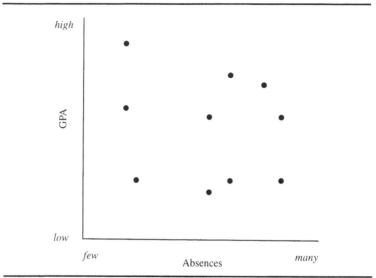

Figure 9.4 Scattergram for a Weak Inverse Relationship

are associated with poor test performance. These people are so anxious that they cannot concentrate on the test. The curvilinear relationship in Figure 9.5 has been exaggerated so that you can see what is meant by *curvilinearity*. Researchers who have found such a relationship have reported

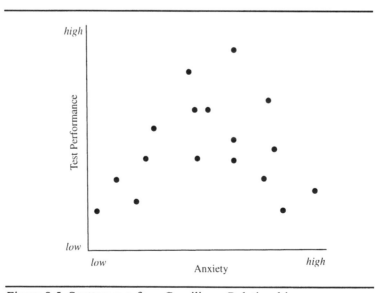

Figure 9.5 Scattergram for a Curvilinear Relationship

many more exceptions in the relationship between anxiety and test performance than are indicated here. Although curvilinear relationships are interesting, they are rarely found in social and behavioral research.

Figures 9.1 through 9.4 illustrate *linear relationships*. We give them this name because the overall pattern of the dots can be described with a straight line even though there may be a great deal of scatter around the line. The vast majority of relationships are *linear*, but, as you now know, some relationships are *curvilinear*.

Pearson *r*

It is important to construct a scattergram to get an overview of the relationship between two sets of scores. When a relationship is linear, we can compute a statistic that concisely describes it. The statistic is the Pearson *r* (also known as the Pearson product-moment correlation coefficient).[1] "Pearson" is the surname of the statistician who developed the statistic, and the letter "*r*" is the name he gave to his statistic. He developed it in such a way that its range is limited to –1.00 to 1.00. Figure 9.6 shows the scattergrams for all four linear relationships we considered above with their corresponding values of *r*. Considering these, you can see some principles. First, when a Pearson *r* is positive in value, it indicates a direct relationship; when it is negative in value, it indicates an inverse relationship. Second, when a Pearson *r* is –1.00 or 1.00, it indicates a perfect relationship (with no scatter on the scattergram). Third, the more scatter there is, the closer the Pearson *r* is to 0.00.

The following chart will help you think about Pearson *r*s and their meanings:

–1.00				0.00				1.00
⇑	⇑	⇑	⇑	⇑	⇑	⇑	⇑	⇑
perfect	strong	moderate	weak	none	weak	moderate	strong	perfect

Notice that precise values of *r* have not been provided for the labels "strong," "moderate," and "weak." This is because researchers do not always agree on the exact interpretation of *r*. One researcher might call a value of .45 "relatively strong," another might call it "moderate," and another might call it "fairly weak."

---

[1] When a relationship is curvilinear, it is inappropriate to use the Pearson *r*. Another statistic, the correlation ratio, which is beyond the scope of this book, should be used.

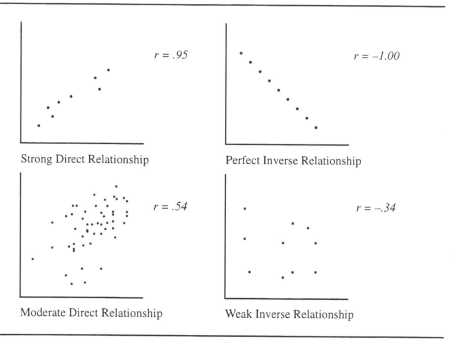

Figure 9.6 Four Scattergrams with Values of the Pearson *r*

Computation of the Pearson *r*

We will consider here the original formula[2] for the Pearson *r*. To use this formula, we must first get the two *z*-scores for each person. You should recall from Chapter 8 that a *z*-score indicates the number of standard deviation units a raw score is from the mean of the group. The formula is:

$$z = \frac{X-M}{SD}$$

We are using *z*-scores because we are often interested in correlating data for two variables expressed on different scales. For example, height seldom exceeds 85 inches while weight rarely dips below 85 pounds for adults. Thus, examining *z*-scores makes the relationship between two variables clearer. To see this, consider again the data on height and weight for

[2]An alternative formula that is easier to use if you are using a calculator instead of a computer is presented in Appendix C. Hence it is called the "computational formula." The one presented in this chapter makes the nature of the correlation coefficient clearer. Hence, it called the "definition formula." The two formulas are algebraically equivalent.

eight people in Table 9.1. The data are repeated here as Table 9.2 with the means and standard deviations shown at the bottom.

Table 9.2 Height and Weight for Eight People

Person	Height (inches)	Weight (pounds)
Jose	72	199
Bill	69	185
Millie	66	190
Kelly	63	155
Frank	62	140
Diana	60	128
Willy	57	132
Jean	54	117
	$M = 62.88$	$M = 155.75$
	$SD = 5.62$	$SD = 29.55$

The raw scores (inches and pounds) in Table 9.2 have been converted to the *z*-scores shown in Table 9.3. Jose's *z*-score for height in Table 9.3 was obtained as follows:

$$z = \frac{X-M}{SD} = \frac{72-62.88}{5.62} = \frac{9.12}{5.62} = 1.62$$

Jose's *z*-score for weight was obtained as follows:

$$z = \frac{X-M}{SD} = \frac{199-155.75}{29.55} = \frac{43.25}{29.55} = 1.46$$

Notice that Jose has very similar *z*-scores on the two variables even though his raw scores of 72 on height and 199 on weight *look* very different from each other. His two *z*-scores are similar because, *relative to the group*, he is both tall and heavy. Examining the *z*-scores for the other people in Table 9.3 indicates that they also have similar *z*-scores on the two variables. Thus, inspection of the *z*-scores indicates that there is a direct relationship (those such as Jose who have high *z*-scores on one variable have high *z*-scores on the other; those such as Jean who have low *z*-scores on one variable have low *z*-scores on the other).

To compute the Pearson *r*, we need to multiply the two *z*-scores for each person and then sum the products. Jose's product is 1.62 x 1.46 = 2.37, which is shown in the last column of Table 9.4. Finally, divide the sum of the products of the *z*-scores (shown at the bottom of the last

Table 9.3 z-Scores on Height and Weight

Person	Height (inches) z-score	Weight (pounds) z-score
Jose	1.62	1.46
Bill	1.09	0.99
Millie	0.56	1.16
Kelly	0.02	−0.03
Frank	−0.16	−0.53
Diana	−0.51	−0.94
Willy	−1.05	−0.80
Jean	−1.58	−1.31

Table 9.4 *z*-Scores on Height and Weight with z-Score Products

Person	Height (inches) z-score	Weight (pounds) z-score	Product of z-scores
Jose	1.62	1.46	2.37
Bill	1.09	0.99	1.08
Millie	0.56	1.16	0.65
Kelly	0.02	−0.03	0.00
Frank	−0.16	−0.53	0.08
Diana	−0.51	−0.94	0.48
Willy	−1.05	−0.80	0.84
Jean	−1.58	−1.31	2.07
			Sum = 7.57

column in Table 9.4) by the number of cases, which is 8. Thus, 7.57/8 = .95, which is the value of *r*, which indicates a very strong relationship between height and weight for this small group of people. (In a larger, more typical group, we would expect to find more exceptions and a lower value of *r* for the relationship between these two variables.)

The formula that describes what we just did is:

$$r = \frac{\sum z_x z_y}{N}$$

Where:

z_x = z-scores on X
z_y = z-scores on Y
N = number of cases

Which variable you call X and which you call Y has no effect on the answer. The formula says to multiply the z-scores on variable X by those on variable Y, sum them, and divide by the number of cases (N).

In the next chapter, we will consider the *coefficient of determination*, which is a statistic that is used when interpreting a Pearson *r*.

A Note on Terminology

The full, formal name of the Pearson *r* is the "Pearson product-moment correlation coefficient." In research reports, you may find it referred to in a variety of ways, such as the "Pearson *r*," the "Pearson correlation coefficient," the "product-moment correlation coefficient," or the "product-moment *r*."

EXERCISE FOR CHAPTER 9

Factual Questions

1. If the dots on a scattergram form a pattern that goes from the lower-left corner to the upper-right corner, what type of relationship is indicated?
   A. direct   B. inverse

2. A lot of scatter among the dots on a scattergram indicates that the relationship is
   A. strong.   B. weak.

3. A Pearson *r* of −.98 indicates that a relationship is
   A. very strong.   B. strong.   C. moderate.   D. weak.   E. very weak.

4. What value of the Pearson *r* indicates the complete absence of a relationship?

5. What value of the Pearson *r* indicates a perfect direct relationship?

6. What is a linear relationship?

7. Draw a scattergram for the raw scores shown in Table 9.5. (Note that the *X*-axis is the horizontal axis and the *Y*-axis is the vertical one.)

Table 9.5 Raw Scores on Math Computations and Reasoning

Person	Math computations raw score *X*	Math reasoning raw score *Y*
Dolly	20	10
Maureen	10	15
Eddie	29	22
Carl	32	28
Bud	39	35
Delleen	40	32
Jean	45	39

8. On the basis of your scattergram, describe the relationship in words (not numbers).

9. Compute the Pearson *r* for the raw scores shown in Table 9.6 on the next page. (Hint: If you use the formula in this chapter, you will first need to compute the *z*-scores using the means and standard deviations that are given in the table. If you use the formula in Appendix C, computation of the *z*-scores will not be necessary.)

Table 9.6 Raw Scores on Vocabulary and Reading

Person	Vocabulary	Reading
Yolanda	40	16
Phyllis	45	20
Marilyn	50	18
Bob	55	22
Ralph	60	24
	$M = 50.00$ $SD = 7.07$	$M = 20.00$ $SD = 2.83$

Questions for Discussion

10. Name two variables other than those mentioned in this chapter between which you would expect to find a direct relationship on a scattergram.

11. Name two variables other than those mentioned in this chapter between which you would expect to find an inverse relationship on a scattergram.

NOTES

Chapter 10

The Coefficient of Determination

In Chapter 9, you learned how to compute a Pearson r and the basics of how to interpret it. You learned that r is expressed on a scale from −1.00 through 0.00 to 1.00. On the surface, a Pearson r *looks like* a proportion because proportions are also expressed on a scale from 0.00 to 1.00. For example, a recent survey indicated that .50 (a proportion) of Americans are opposed to affirmative action. A proportion may be converted to a percentage by multiplying by 100. Thus, the proportion .50 x 100 = 50%. However, a Pearson r is *not* a proportion; therefore, multiplying it by 100 does *not* yield a percentage of anything. For example, a Pearson r of .50 is *not* equivalent to 50%.

Let's consider this problem by using an analogy. Suppose that Mr. Pearson had devised the one-foot ruler using the same mathematical principle that he used to invent r. This is what his one-foot ruler would have looked like with the inches indicated on it:

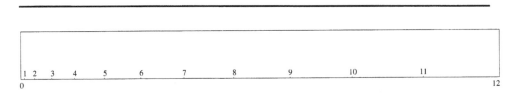

Figure 10.1 Pearson's Hypothetical Ruler for Measuring Inches

You might have expected Mr. Pearson to place the 6 halfway between 0 and 12, but, as you can see in Figure 10.1, he placed it 1/4 of the way up from 0. Clearly, Mr. Pearson's ruler is a non-equal interval instrument (that is, the intervals among the numerals are not equal in size). Such an instrument has much potential for confusion when you consider that 3 is not half of six on the ruler, 4 is not half of 8, and so on.

If relationships could be measured with a ruler-like instrument marked with values of r instead of inches, this is what Mr. Pearson's instrument

71

would look like for direct (positive) relationships with correlation coefficients indicated on it:

Figure 10.2 Pearson's Hypothetical Ruler for Measuring Correlation

As you can see in Figure 10.2, an *r* of .50 is *not* halfway between 0.00 and 1.00. In fact, it is 1/4 of the way up from 0.00. What does this mean? Simply that a Pearson *r* of .50 is 1/4 or 25% higher than 0.00. For example, if we build a basic math test that we hope will predict success in a first-semester algebra course, we could administer the math test before the students take algebra and, at the end of the semester, administer an algebra achievement test. If we correlated these two tests and got a Pearson *r* of 0.00, it would mean that the math test has no ability to predict algebra achievement; if we got 1.00, it would mean that the math test is a perfect predictor. However, suppose we got a Pearson *r* of .50, which is about what many researchers have gotten when conducting this type of study. The value of .50 means that the math test is 25% better than no ability to predict. In other words, it is 25% better than 0.00 — *not* 50% better. Of course, this seems to defy logic, but remember Pearson's ruler: the values were not placed where you might think they logically should be.

Clearly, the interpretation of the Pearson *r* is muddied by its non-equal interval scale. We can get around this problem rather simply by calculating and using a related statistic that does not have this problem: the *coefficient of determination*. The calculation is quite simple: just square *r*. Thus, for a Pearson *r* of .50, the coefficient of determination is .50 x .50 = .25. Other coefficients of determination for selected values of *r* and the associated percentages are shown in Table 10.1. Keep in mind that the percentages tell you how much better off you are with selected values of *r* than you would be with a value of 0.00. In other words, with a value of 0.00 there is no correlation between two variables. For example, a Pearson *r* of .10, according to Table 10.1, means that you are only 1% better off than having a Pearson *r* of 0.00.

Table 10.1   Coefficients of Determination for Selected
Values of r

Pearson r	Coefficient of Determination (r^2)	Percentage
.00	.00	0%
.10	.01	1%
.20	.04	4%
.30	.09	9%
.40	.16	16%
.50	.25	25%
.60	.36	36%
.70	.49	49%
.80	.64	64%
.90	.81	81%
1.00	1.00	100%

Notice in Table 10.1 that smaller values of r shrink more dramatically when squared than larger values do. Thus, we should be especially careful when interpreting small values of r.

Of course, you can get the value of the coefficient of determination that corresponds to any value of Pearson r simply by squaring the Pearson r. Note that the symbol for the coefficient of determination is r^2, which tells you how to compute it. To get the percentage, just multiply by 100.

The percentage that corresponds to the coefficient of determination has a special name — the *explained variance* (also known as *variance accounted for*). So if you're writing a research report, you could report that r = .30, r^2 = .09, and that the *explained variance* is 9%.

By now you may be wondering why we bother reporting the Pearson r at all since it is flawed and since there is an alternative—the coefficient of determination. The reason we report it is that all coefficients of determination are positive values (for example, if $r = -.80$, $r^2 = .64$). Since the difference between direct and inverse relationships is very important, the coefficients of determination is deficient because it fails to differentiate between them. Thus, if we reported coefficients of determination without the associated Pearson rs, in each case we would need to explain whether they represent direct or inverse relationships. Furthermore, some people might look only at the coefficients and miss our explanations regarding the directions of the relationships. Thus, it is best to report both the Pearson r, to indicate the direction of the relationship, and the coefficient of determination, to help in the interpretation of the size of the relationship.

In practice, many applied researchers report only the values of the Pearson rs when describing correlation and not the values of the coefficients of determination. When you encounter this in research reports, it is important to calculate the corresponding values of r^2 (even if only by mental estimation) in order to interpret the values of r.

EXERCISE FOR CHAPTER 10

Factual Questions

1. Should a Pearson r be thought of as a proportion?

2. Expressed as a percentage, how much better than 0.00 is a Pearson r of .50?

3. Which values of r shrink more dramatically when squared?
   A. small values   B. large values

4. What is the symbol for the coefficient of determination?

5. What is the value of the coefficient of determination for a Pearson r of .65?

6. For a coefficient of determination of .75, what is the corresponding percentage of explained variance?

7. According to this chapter, which of the following should you report when describing correlation?
   A. Pearson r only
   B. coefficient of determination only
   C. both the Pearson r and the coefficient of determination

Question for Discussion

8. Suppose you located a reading readiness test (to be given during kindergarten) that had a Pearson *r* of .40 when correlated with reading achievement measured at the end of first grade. What is the value of the coefficient of determination? On the basis of this, do you think that the reading readiness test is a good predictor of reading achievement? Explain.

NOTES

Chapter 11

Linear Regression

At typical colleges and universities, we often find that scores on a college admissions test, such as the College Board's SAT-Verbal, correlate modestly with freshman GPAs in college. We would not be surprised to get a Pearson r of .45. This translates into a coefficient of determination of .20, which means that an r of .45 is 20% better than an r of 0.00, which is what we'd get if we pulled names out of a hat to decide whom to admit to our university.[1] Now let's suppose that we decide that 20% is good enough to warrant using the test for selecting among the applicants who apply to our university. How can we use this test to make decisions about *individual* applicants? (Note that Pearson rs and coefficients of determinatio tell us only about *trends across groups*.) The answer is to use the technique called *linear regression*.

To use linear regression to make predictions for individuals, we first have to administer the predictor (in this case, the SAT-Verbal) to a group of applicants, admit them to our university, and then gather their freshman GPAs. At this point, we have two scores for each student (SAT-Verbal and GPA). Using these data to calculate a formula for linear regression, we can then make predictions about future applicants to our university.

As you know from Chapter 9, the term *linear* refers to a single straight line. Thus, linear regression is a technique in which we use a straight line to make predictions for individuals. Before examining how linear regression works, let's first review the equation for a straight line, which you probably first encountered in elementary geometry:

$$Y = a + bX$$

Where:

Y = score on variable Y

a = intercept

b = slope

X = score on variable X

---

[1] See Chapter 9 to review the Pearson r, and Chapter 10 to review the coefficient of determination.

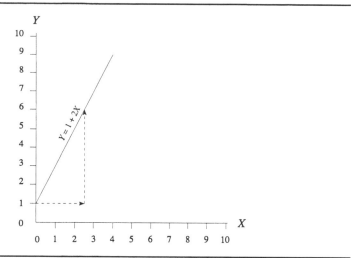

Figure 11.1 Straight Line Defined by $Y = 1 + 2X$

Figure 11.1 shows the straight line described by this equation:

$$Y = 1 + 2X$$

Notice that the straight line starts at a score of 1, which is the *intercept*. In other words, the intercept is the point where the line begins on the Y-axis. The dashed lines show that for every one unit we move to the right, we have to go up two units to reach the line. A unit can be an inch, a foot, or a mile! Whatever the unit is, the *slope* of 2 tells us we must go up 2 units to reach the line. Thus, you can see that the slope is the rate of change in the line. In Figure 11.1, the dashed line going to the right from the Y-axis is 1/2 inch long and the dashed line going up to the solid line is 2 times that: 1 inch long. (To avoid confusion, be careful to note that the slope does *not* refer directly to the scores on the X or Y axis. Instead, it refers to the relationship between the two dashed lines.)

Since some students have difficulty in understanding the slope, let's consider a line with a different slope. Figure 11.2 also has an intercept of 1 but has a slope of .5 (one-quarter of the slope in Figure 11.1), as indicated by this equation:

$$Y = 1 + .5X$$

78

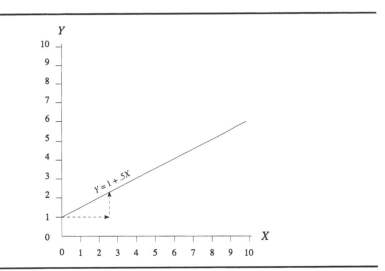

Figure 11.2   Straight Line Defined by $Y = 1 + .5X$

Because the slope is smaller in Figure 11.2 than in Figure 11.1, the line in Figure 11.2 rises more slowly than the one in Figure 11.1. In other words, Figure 11.1 has a steeper slope than Figure 11.2.

Why Use a Straight Line for Prediction?

You probably recall from Chapter 9 that most relationships are linear; that is, when plotted on a scattergram, the dots form a pattern that could be reasonably described by a single straight line. For example, consider Figure 11.3, which shows a scattergram for the scores in Table 11.1. To keep our computations simple later on, we're using small numbers for the predictor of college success (such as the College Board's tests) and for the scores on the outcome in college (such as GPAs). The scattergram in Figure 11.3 shows a direct linear relationship that could be described by a straight line from the lower-left to the upper-right. If we drew a straight line to describe the pattern formed by the dots, we could then use the line to predict scores on variable Y from scores on variable X. Before we do this to see how it works, we first need to determine *exactly* where to draw the line. In other words, we don't want just any old line but, rather, the one that best describes the relationship. The computational procedures we'll consider below determine the position of this best-fitting line.

Table 11.1 Scores on a Predictor and
Outcome Variable

Student	Predictor X	Outcome Y
Bill	5	8
Jane	7	7
Tim	2	1
Saul	1	1
Fernando	4	3
Lester	2	3

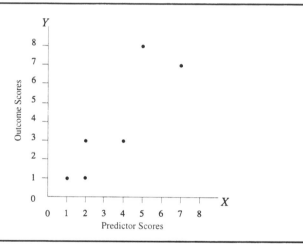

Figure 11.3 Scattergram for the Scores in Table 11.1

Computational Procedures

The formulas for linear regression will be illustrated using the data in Table 11.1, which are repeated in Table 11.2. The values of X in column 2 are the predictor scores (such as College Board scores), and the values of Y in column 3 are the outcome scores (such as college GPAs).

First, we need to compute the values in columns 4 and 5 of Table 11.2. To get column 4, square the values of X (such as 5 x 5 = 25 for Bill), then sum the column to get 99 at the bottom of the column. Next, multiply each

Table 11.2 Worktable for Linear Regression

Col. 1	Col. 2	Col. 3	Col. 4	Col. 5
Student	X	Y	X^2	XY
Bill	5	8	25	40
Jane	7	7	49	49
Tim	2	1	4	2
Saul	1	1	1	1
Fernando	4	3	16	12
Lester	2	3	4	6
Sums	21	23	99	110

value of X by the corresponding value of Y to get the values in column 5. For example, for Bill, 5 x 8 = 40. The sum of column 5 is 110.

With the sums from a worktable such as Table 11.2, we are ready to calculate the slope (b) using this formula:

$$b = \frac{\sum XY - [(\sum X)(\sum Y) \div N]}{\sum X^2 - [(\sum X)^2 \div N]}$$

Where:

$\sum XY$ = sum of column 5
$\sum X$ = sum of column 2
$\sum Y$ = sum of column 3
$\sum X^2$ = sum of column 4
N = number of cases (students)

Solving for the data we are considering, we get:

$$b = \frac{110 - [(21)(23) \div 6]}{99 - [21^2 \div 6]} = \frac{29.5}{25.5} = 1.157$$

Next, we calculate the value of the intercept (a) using this formula:

$$a = M_Y - bM_X$$

Where:

M_Y = mean of the scores on Y
b = value of the slope that we just computed
M_Y = mean of the scores on X

It turns out that the mean of Y is 3.833 and the mean of X is 3.500. Thus:

$$a = 3.833 - (1.157)(3.500) = -0.217$$

Now, for the data in Table 11.1, we have this equation:

$$Y = -.217 + 1.157X$$

Figure 11.4 shows the equation plotted on the scatter diagram for the data we are considering. Notice that the intercept happens to be a small negative number, so the line starts slightly below zero.

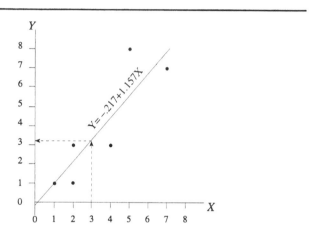

Figure 11.4 Scattergram for the Scores in Table 11.1 with the Best-fitting Straight Line

Now let's suppose that a new person comes along and we administer Test X to her, and she obtains a score of 3. What is her predicted score on Y? We can get the predicted score by referring to the line on the scattergram. We do this by looking up a score on X of 3, go up to the solid line we've drawn and go left to the Y-axis, where we find a score slightly higher than 3. The dashed lines in Figure 11.4 show how to go up from X and over to Y.

Instead of using the scattergram to get a predicted score, we can calculate it by substituting 3 for X in the equation for a straight line with the values of a (−.217) and b (1.157) that we calculated above:

$$Y = -.217 + (1.157)(3) = -.217 + 3.471 = 3.254 = 3.3$$

This is essentially the same answer that we got by referring to the line on the scattergram. In academic journals, researchers are more likely to give you the values of a and b than they are to give you a scattergram with the best-fitting line drawn in, so it's important to learn how to use the values to make predictions.

To review: as a practical matter, all we need to use linear regression is to calculate the values of a and b based on the performance of a previous group. When a new individual comes along, we substitute her score on the predictor for X in the equation and solve for Y, which is the predicted score. Let's consider this equation for predicting GPAs from SAT-Verbal scores, which range from 200 to 800, based on the performance of a large group at one particular college:

$$Y = -2.000 + .008X$$

For a student with an SAT of 300, the predicted GPA is:

$$Y = -2.000 + (.008)(300) = 0.4$$

For a student with an SAT of 500, the predicted GPA is:

$$Y = -2.000 + (.008)(500) = 2.0$$

For a student with an SAT of 700, the predicted GPA is:

$$Y = -2.000 + (.008)(700) = 3.6$$

As you can see, these predictions make sense. Those with low SAT scores have low predicted GPAs. Those with high SAT scores have high GPAs.

Accuracy of the Predictions

How good are the predictions we make with linear regression? How much confidence can we have in them? While there are rather precise methods for answering these questions that are beyond the scope of this book, we can get a handle on it by thinking about the Pearson r and coefficient of determination. If the Pearson r for the relationship between X and Y equals 1.00, we have a perfect correlation and our predictions should be perfect. The lower the Pearson r, the more errors we will make when we predict scores on Y from scores on X. As it turns out, the value of the Pearson r for the data in Table 11.1 is .87, which is strong. The corresponding coefficient of determination is .76 (about 76% better than no ability to predict), which is quite high considering that we are trying to predict complex

83

human behavior (such as earning grades in college). Thus, in the example we're considering, our prediction system should work very well.

In practice, when we try to predict grades in college with an admissions test, we're fortunate to get a correlation coefficient as high as .45. As we noted earlier in this chapter, an r of .45 translates into a coefficient of determination of .20 (about 20% better than no ability to predict). While most researchers consider this considerably better than nothing, it means that there will be many errors. Thus, many applicants who are admitted on the basis of linear regression will not do well in college, and many who are not admitted would have succeeded if they had been admitted. Still, with a coefficient of determination of .20, we're 20% better off than pulling names out of a hat to make admissions decisions.

We can increase our ability to predict by using more than one predictor. For example, we could use both the College Board's SAT-Verbal and SAT-Math as well as high school GPAs to predict grades in college. Using these three predictors instead of just one usually increases our ability to predict very modestly. The mathematics of using more than one predictor is beyond the scope of an introductory text.

EXERCISE FOR CHAPTER 11

Factual Questions

1. In the equation for a straight line, for what does the letter a stand?

2. Using the formula $Y = 2.34 + .07X$, what is the predicted score on Y for a person who has a score of 400 on X?

3. What does *linear* mean?

4. What are the values of a and b for the scores in Table 11.3?

Table 11.3   Scores for Question 4

Subject	Score on X	Score on Y
Carol	2	0
Robert	1	3
Rebecca	4	5
Lynn	7	8
Hubert	6	9

5. Using the values of a and b that you calculated for question 4, what is the predicted value of Y for a new person who has a score of 3 on X?

6. Using the values of a and b that you calculated for question 4, what is the predicted value of Y for a new person who has a score of 5 on X?

7. Using the values of a and b that you calculated for question 4, what is the predicted value of Y for a new person who has a score of 4 on X?

Question for Discussion

8. Name a predictor and an outcome variable studied in your professional field that might be subjected to linear regression in order to make predictions for individuals.

NOTES

Chapter 12

Glossary of Descriptive Statistics

This chapter contains definitions of the major terms you have encountered in this book. You may use it for review by considering each term in bold, seeing if you can define it, and then reading the definition to see if your definition is correct. Since the definitions given here are brief, page numbers are provided where you may find more detailed definitions and related information.

CHAPTER 1

Population The group of interest to a researcher. It may be small, such as all Girl Scouts in a troop, or large, such as all Girl Scouts in the United States. (pp. 2–3)

Sample A group that is studied in order to make inferences about the population from which it was drawn. (p. 2)

Inferential statistics The type of statistics that are used to make inferences from samples to populations. (pp. *v*, 2)

Descriptive statistics The type of statistics that are used to organize and summarize data. (p. 2)

Scales of measurement A classification system for describing levels of measurement: nominal, ordinal, and equal interval. (pp. 3–4)

Nominal scale of measurement Type of measurement in which individuals are classified with words such as "male" or "female." (pp. 3–4)

Ordinal scale of measurement Type of measurement in which individuals are classified according to their rank order. (pp. 3–4)

Equal interval scale of measurement Type of measurement in which the points along the scale have equal intervals among them, such as inches on a ruler. (pp. 3–4)

CHAPTER 2

Percentage A value expressing the rate per 100. Thus, if 52% of the people in a population are women, 52 out of every 100 are women. (p. 7–8)

Bar graph A statistical figure for nominal data that is made up of bars. Usually, the bars show the number or percentage of individuals in each category of nominal data. (p. 9)

CHAPTER 3

Frequency distribution A statistical table in which scores are listed in the first column, frequencies in the second, and, sometimes, percentages are listed in column(s) to the right. (pp. 13–15)

Polygon A statistical figure or drawing that depicts a frequency distribution with lines that form a curve. With small data sets, the curves are often jagged. The horizontal axis is labeled with scores, and the vertical axis is labeled with frequencies or percentages. Also known as a *curve*. (p. 15)

Normal curve The most common curve (polygon). Has a symmetrical bell shape. (pp. 15–16)

Skewed curve A curve (polygon) that has one tail longer than the other. (pp. 15–16)

Skewed to the right A curve (polygon) having a longer tail to the right than to the left. The tail to the right is created by a relatively small number of cases that have very high scores. Also known as a *positive skew*. (p. 16)

Skewed to the left A curve (polygon) having a longer tail to the left than to the right. The tail to the left is created by a relatively small number of cases that have very low scores. Also known as a *negative skew*. (pp. 16–17)

CHAPTER 4

Mean The balance point in a distribution. It is the most frequently used average. Obtained by summing the scores and dividing by the number of cases. Inappropriate for use with highly skewed distributions. Appropriate only for equal interval data. (pp. 21–24)

CHAPTER 5

Median The average that is defined as the middle score in a distribution. Has half the cases above it and half below it. Used instead of the mean when distributions are highly skewed and for obtaining the mean for ordinal data. (pp. 27–28)

Mode The average that is defined as the most frequently occurring score. Rarely reported in scientific writing. (p. 29)

CHAPTER 6

Variability General term for referring to variation in scores (that is, differences among scores). Synonyms are *spread* and *dispersion*. (p. 33)

Standard deviation The most popular measure of variability. Based on the differences of the individual scores from the mean of the group. Usually reported in conjunction with the mean. Has a special relationship to the normal curve. (pp. 33–40)

68% rule In a normal distribution, about 68% of the cases lie between the mean and one standard deviation unit on both sides of the mean. (p. 37–40)

95% rule In a normal distribution, about 95% of the cases lie between the mean and two standard deviation units on both sides of the mean. (pp. 37–39)

99.7% rule In a normal distribution, 99.7% of the cases lie between the mean and three standard deviation units on both sides of the mean. (pp. 38–39)

CHAPTER 7

Range A measure of variability that indicates the number of score points covered from the highest score to the lowest score. Seldom used in scientific reporting. (pp. 43–44)

Quartiles A measure of variability that consists of the three score values that divide a distribution into quarters. The *first quartile* is the score with 25% of the cases below it. The *second quartile* is the score with 50% of the cases below it. The *third quartile* is the score with 75% of the cases below it. (pp. 45–46)

Interquartile range The range of the middle 50% of the subjects (that is, between the first quartile and the third quartile). A measure of variability that is reported when the median is reported as the average. (pp. 46–47)

CHAPTER 8

Raw score The number of points an examinee earns on a test or measure. (p. 49)

Standard score A score that indicates how many standard deviation units an examinee's score is from the mean of a norm group. A negative standard score indicates that an examinee is below the mean; a positive one indicates that he or she is above the mean. Standard scores are also known as *z*-scores. (pp. 49–55)

***z*-score** See *standard score*.

Transformed standard scores Standard scores or *z*-scores that have been transformed to have a new mean and a new standard deviation. (pp. 51–55)

CEEB scores Transformed standard scores with a mean of 500 and a standard deviation of 100. The acronym *CEEB* stands for the College Entrance Examination Board. (pp. 51–52)

IQ scores Transformed standard scores with a mean of 100 and a standard deviation of 15. The acronym *IQ* stands for Intelligence Quotient. (pp. 52–53)

McCall's *T* scores Transformed standard scores with a mean of 50 and a standard deviation of 10. Also referred to simply as *T* scores. (p. 53–54)

CHAPTER 9

Scattergram A statistical figure that uses dots on a graph to show the relationship between two quantitative variables. When the dots form a pattern from the lower-left to the upper-right, the relationship is direct. When the dots form a pattern from the upper-left to the lower-right, the relationship is inverse. The more scatter, the weaker the relationship. Also known as a *scatter diagram* or *scatterplot*. (pp. 57–64)

Pearson *r* A statistic that summarizes in a single number the relationship between two quantitative variables. It ranges from 0.00 (no relationship) to 1.00 (perfect direct relationship) and from 0.00 to −1.00 (perfect inverse relationship). (pp. 63–67)

Correlation coefficient See *Pearson r*.

CHAPTER 10

Coefficient of determination A statistic that is used to interpret a Pearson *r*. Computed by squaring *r*, its symbol is r^2. When multiplied by 100, the coefficient of determination indicates the percentage better than 0% that the corresponding Pearson *r* represents. (pp. 71–74)

CHAPTER 11

Linear regression A technique used to make predictions from one quantitative variable to another using a straight line based on the performance of some group on the two variables. (pp. 77–84)

Intercept The value where a straight line intersects with the Y-axis (vertical axis). The symbol is a. Calculated in order to use linear regression. (pp. 77–83)

Slope The rate of change in a straight line used in linear regression. Its symbol is b. (pp. 77–83)

Appendix A

Formula for the Interpolated Median

In Chapter 5, you learned how to compute the median by putting the scores in order from low to high and counting to the middle to find the median. When there are ties in the middle, an interpolated median can be computed. For a given set of data, the median you learned how to compute in Chapter 5 will be very close to the interpolated median that you will learn how to compute in this appendix. In fact, sometimes the two will be identical. Nevertheless, some consider the interpolated median to be more precise that the median obtained by simple counting when there are ties in the middle of the distribution.

In the distribution shown in Table A.1, the middle of the distribution is at a score of 17 because about half the cases are above 17 and half are below. (See Chapter 3 to review frequency distributions.) A column has been added for "cumulative frequency," which we will need for the computation of the interpolated median. It has been obtained by summing the frequencies in the second column from the bottom up. For example, the bottom score interval is for a score of 14, which has a frequency of 2. The

Table A.1 Frequency Distribution for Appendix A

Score	Frequency (f)	Cumulative frequency (cf)
20	2	22
19	0	20
18	4	20
17	6	16
16	5	10
15	3	5
14	2	2
	N = 22	

cumulative frequency is the frequency of 2 plus all those below it; since there are no frequencies below it, the bottom cumulative frequency is $2 + 0 = 2$, which is the bottom entry in the last column. The cumulative frequency associated with the score of 15 is obtained by summing the frequency associated with it with the frequency for the score below it ($3 + 2 = 5$). The cumulative frequency associated with the score of 16 is obtained by summing the frequencies for scores of 16, 15, and 14 ($5 + 3 + 2 = 10$). The cumulative frequency associated with the score of 17 is obtained by summing the frequencies for scores of 17, 16, 15, and 14. An analogous procedure has been followed for the other scores. As you can see, the cumulative frequencies show the total number of cases in and below any given score interval.

The formula for the interpolated median is:

$$\text{Median} = L + \frac{N(.5) - cfb}{f}$$

Where:

L = exact lower limit of the interval that contains the median; this is the *interval of interest*. To find this interval, divide N by 2 (in this case, $22/2 = 11$), and go *up* the *cf* column looking for subject number 11. The *cf* column indicates that up to a score of 16, there are 10 subjects. The next score interval (17) contains subjects 11 through 16. Since we are looking for subject 11, **17** is the score interval of interest. Subtract .5 from the score of interest to obtain the exact lower limit (i.e., $17 - .5 = \mathbf{16.5}$).

N = total number of cases; in this example, it is **22**.

cfb = cumulative frequency (*cf*) for the score immediately *below* the *interval of interest* (in this case, $cfb = \mathbf{10}$ for the score of 16, which is the score below 16.5).

f = frequency of the interval of interest (in this case, $f = \mathbf{6}$ for the interval associated with a score of 17).

Substituting into the formula, we obtain the following:

$$\text{Median} = 16.5 + \frac{(22)(.5) - 10}{6} = 16.5 + \frac{1}{6} = 16.5 + .167 = 16.667 = 16.67$$

Notice that by using the method in Chapter 5, we would have counted to the middle and obtained a median of 17, which is very close to 16.67. When rounded to a whole number, the two methods give the same answer for these data.

Appendix B

Computational Formula for the Standard Deviation

In Chapter 6, you learned how to compute the standard deviation using a formula that clearly shows that the size of the standard deviation is influenced by the amount by which the individual scores deviate from the mean of a group. An algebraically equivalent formula that is easier to use is presented in this appendix. The formula is:

$$SD = \frac{1}{N}\sqrt{N\Sigma X^2 - (\Sigma X)^2}$$

Where:

N = number of cases

ΣX^2 = sum of the squared scores
(sum after squaring each score)

$(\Sigma X)^2$ = square of the sum of the scores
(sum and then square the sum)

Use of the formula will be illustrated with the scores in the first column of Table B.1. Notice that in the second column, the scores have been squared. Both columns have been summed.

Table B.1 Scores for Appendix B

X	X^2
3	9
5	25
7	49
8	64
$\Sigma X = 23$	$\Sigma X^2 = 147$

$$SD = \frac{1}{4}\sqrt{(4)(147) - (23)^2} = .25\sqrt{588 - 529} = (.25)(7.681) = 1.92$$

Appendix C

Computational Formula for the Pearson *r*

In Chapter 9 you learned how to compute the Pearson *r* using a formula that clearly shows that it is based on *z*-scores. In this appendix, you will learn how to use a formula that is algebraically equivalent but easier to use computationally. The formula is:

$$r = \frac{N\Sigma XY - (\Sigma X)(\Sigma Y)}{\sqrt{[N\Sigma X^2 - (\Sigma X)^2][N\Sigma Y^2 - (\Sigma Y)^2]}}$$

To use the formula, first set up a worktable such as the one in Table C.1. One set of scores has been designated as *X* and the other as *Y*. It is customary to designate the one measured first as *X* and the other one as *Y*. However, you will get the same answer either way.

Table C.1 Worktable for Computing *r*

Col. 1	Col. 2	Col. 3	Col. 4	Col. 5	Col. 6
Student	X	Y	X^2	Y^2	XY
Bill	5	0	25	0	0
Liz	7	4	49	16	28
Carol	0	9	0	81	0
Stu	1	7	1	49	7
Frank	4	5	16	25	20
Brandy	2	6	4	36	12
$\Sigma =$	19	31	95	207	67

In Table C.1, the scores are shown in columns 2 and 3. Begin your work by completing the worktable as follows. Compute the values in column 4 by squaring each value of *X*. Compute the values in column 5 by

95

squaring each value of Y. Finally, compute the values in column 6 by multiplying each X by each Y. (For example, Brandy's score of 2 on X is multiplied by her score of 6 on Y to obtain the product of 12 in column 6.) Then sum the values in columns 2 through 6.

The formula presented at the beginning of this section is rewritten below using column numbers to refer to the *sums* of the columns shown in Table C.1.

$$r = \frac{N(Col.6)-(Col.2)(Col.3)}{\sqrt{[N(Col.4)-(Col.2)^2][N(Col.5)-(Col.3)^2]}}$$

$$r = \frac{6(67)-(19)(31)}{\sqrt{[(6)(95)-19^2][(6)(207)-31^2]}} = \frac{-187}{\sqrt{[209][281]}} = \frac{-187}{242.341} = -.772 = -.77$$

Index

Bar graph 9-10
Caption 8
CEEB score 51-52
Central tendency 29
Coefficient of determination 71-74
Correlation coefficient 67
Curve 15-17
Curvilinear relationship 61-63
Deviation score 21-22
Direct relationship 58-59
Dispersion 33
Equal interval measurement 3-4, 13, 23
Explained variance 73
Five-point description 46-47
Frequency distribution 13-15
Histogram 17-18
Inferential statistics 2
Instruments 3
Intercept 77-83
Interquartile range 46-47
Interval scale of measurement 4
Inverse relationship 60-61
IQ score 52-53
Linear regression 77-84
Linear relationship 63, 79
McCall's T score 53
Mean 21-24, 49-55
Median 27-28, 45-47, 91-92
Mode 29
Nominal scale of measurement 3-4, 23
Normal curve 15-16, 36-40

Ordinal scale of measurement 3-4, 23
Outlier 23
Parameter 2
Pearson r 63-67, 83-84, 95-96
Percentages 7-10
Polygon 15-17
Population 2-3
Product-moment correlation coefficient 67
Quartiles 44-46
Range 43-44
Ratio scale of measurement 4
Rules, for statistical table 8
Sample 2
Scales of measurement 3-4
Scatter diagram 57
Scattergram 57-64
Scatterplot 57
Score interval 14
Score, symbol for 14
Skewed curve 16-17
Slope 77-83
Spread 33
Standard deviation 33-40, 49-55, 93
Standard score 49-55
T score 53
Table, statistical 8
Transformed standard score 51-55
Variability 33
Variance accounted for 73
z-score 49-55, 64-67